PROGRESS IN BIOMETEOROLOGY, Vol. 2
EFFECTS OF SHELTER ON THE PHYSIOLOGY OF PLANTS AND ANIMALS

PROGRESS IN BIOMETEOROLOGY

EDITOR H. LIETH   FOUNDER: S.W. TROMP

VOLUME 2

SWETS & ZEITLINGER B.V., LISSE 1985

Effects of Shelter on the Physiology of Plants and Animals

Edited by John Grace

SWETS & ZEITLINGER B.V., LISSE 1985

CIP-GEGEVENS KONINKLIJKE BIBLIOTHEEK, DEN HAAG

Effects

Effects of shelter on the physiology of plants and animals
/ ed. by John Grace. - Lisse : Swets & Zeitlinger. - Ill.
- (Progress in biometeorology ; vol. 2)
Met lit. opg.
ISBN 90-265-0603-1 geb.
SISO 556 UDC 551.58:[591.1+581.1]
Trefw.: plantenfysiologie ; biometeorologie / fysiologie
der dieren ; biometeorologie.

ISBN 90 265 0603 1

Printed in the Netherlands by Offsetdrukkerij Kanters B.V., Alblasserdam

CONTENTS

PREFACE

The importance of natural and man-made shelter has been recognised for centuries. Broadly speaking, shelter is the amelioration of the local climate. From a biophysical viewpoint, the influence of any form of shelter on the conditions for life is inherently complex and variable. Landmarks in the scientific study of shelter include the contributions by Bates (1911), Geiger (1960) and Jensen (1954). More recently, physical principles have been applied to the energy balance of plants and animals, first by Gates (1962) and Monteith (1973), and then by many others. Also, appropriate instrumentation has become more widely available (see Fritschen & Gay, 1979). Thus, we are approaching a stage at which a useful understanding of the influence of shelter on heat and mass transfer between organisms and their environment is within our grasp. This is not to say that we know all. The aerodynamic properties of biological structures and of the roughness elements in the landscape are extremely complex and merit much further study.

The biological responses of organisms to shelter are variable. In the case of plants much new work has accumulated in the last fifteen years or so. This has forced a re-evaluation of the role of shelter in evaporation, and has gone some way towards understanding the influence of shelter on growth. Two new phenomena have come to light: the abrasion of leaf surfaces, which may influence surface resistances; and the influence of mechanical action *per se* on plant development. In the case of animals, there has been progress on the modelling of heat transfer and further advances in the characterisation of the stressed condition (Stanier, Mount & Bligh, 1984).

The chapters of this book arose from a meeting of working scientists whose common interest was in shelter and its effects. Their viewpoints and approaches were, however, quite different; thus giving rise to what I hope the reader will agree is an interesting and useful book.

Participants were drawn from three groups: the Environmental Physiology Group of the Society for Experimental Biology, the Agricultural Meteorology Group of the Royal Meteorological Society, and the Climatic Physiology Group. The meeting was held at the University of Hull, April

1983, as part of the regular gathering of the Society for Experimental Biology.

I have to thank those who helped me to organise the meeting, especially Mike Dennett, Ernest Hamley and Lynton Incoll. Sessions were introduced and chaired by Professors J.L. Monteith, L.E. Mount and P.G. Jarvis.

In the production of this book I have been assisted by Anne Cuthbertson and Pam Armstrong. Drawings were contributed by Jean Lothian.

References

Bates, C.G. (1911). Windbreaks: their influence and value. U.S. Department of Agriculture, Forest Service Bulletin 86. Government Printing Office, Washington.

Fritschen, L.J. & Gay, L.W. (1979). Environmental Instrumentation. Springer-Verlag: New York.

Gates, D.M. (1962). Energy Exchange in the Biosphere. Harper & Row: New York.

Geiger, R. (1960). The Climate near the Ground. Harvard University Press: Massachusetts.

Jensen, M. (1954). Shelter Effect. Danish Technical Press: Copenhagen.

Monteith, J.L. (1973). Principles of Environmental Physics. Arnold: London.

Stanier, M.W., Mount, L.E. & Bligh, J. (1984). Energy Balance and Temperature Regulation. Cambridge University Press: Cambridge.

J. Grace,
Edinburgh, 1984.

Progress in Biometeorology, 1985, Vol. 2, pp. 1-16

CONVECTIVE HEAT TRANSFER FROM LEAVES

J. Grace & M. Dixon
Department of Forestry & Natural Resources,
University of Edinburgh,
Mayfield Road,
Edinburgh, EH9 3JU.

INTRODUCTION

The fundamental effect of shelter from the wind is to reduce the air flow around the organism and thus influence the rates of transfer of heat and material that occur by convection between the organism and the environment. In this sense, an understanding of convective heat transfer is crucial to any interpretation of the shelter effect on plants and animals. Other, biological, effects follow from these primary, physical, influences.

Calculation of this shelter effect is not straightforward, as the aerodynamics of natural surfaces are complicated, especially when temperature differences between surfaces and the atmosphere give rise to an additional vertical component of air movement through bouyancy.

The transport of heat from a surface to a moving fluid is termed convection. Although this chapter is concerned almost exclusively with convection, other modes of heat transfer are involved in most practical cases of shelter. It should be realised that conduction is associated with convective heat transfer in two important ways. Firstly, at the surface itself, where the air is at rest, the transport of heat from the surface to the air is by conduction. In the case of hairy leaves or deeply folded developing leaves this zone of conduction may be considerably extended. Secondly, the maintenance of local surface-to-air gradients of temperature may depend crucially on conduction of heat within the plant tissue, especially in bulky tissues like those of many important crops. Radiative heat transfer also has a role to play as the environment is often far from isothermal. Moreover, out-of-doors there is usually a strong component of short-wave radiation from the sky. Furthermore, shelter belts and other structures inevitably alter the radiation regime: so much so that the term 'shelter' in animal studies frequently implies radiative shelter (see Gatenby, in a later chapter). This is not however, generally the case in studies of plants, where 'shelter' generally means shelter from the wind, and only very close to the shelter is it necessary to consider effects of shade.

The convective mode of heat transfer can be divided into two basic types. When the air flow is being driven from outside the system under study, as when wind blows on a leaf, the process is called forced convection. If on the other hand, the air flow is generated by thermal gradients within the system under study, as when a leaf in bright sunlight warms on a windless day, the process is called natural, or free, convection. In the former case the flows are generally more vigorous and the transfer rates are generally higher, at least for the conditions prevailing for biological subjects at ground level on Earth. As we will see later, there are many cases in which both forms of convection are important. Such a regime is called mixed, or hybrid, convection.

TERMINOLOGY

There are several ways in which the susceptibility of a surface to convective heat loss may be expressed. All have been used in the plant sciences. The heat transfer coefficient, h is the oldest though perhaps the least useful:

$$h = C/(T_s - T) \qquad 1)$$

where C is the rate of convective heat transfer per unit area (W m^{-2}), T_s is the surface temperature and T is the air temperature. The numerical value of the heat transfer coefficient (W m^{-2} °C^{-1}) is found to depend mainly on the area and shape of the surface and the air speed. The relationship appears to have first been suggested by Newton as an expression of the fact that heat transfer by convection is primarily a diffusive process, occuring from hot to cold at a rate proportional to the magnitude of the temperature difference between the body and the surrounding fluid.

Plant physiologists have for a long time used electrical analogies to represent diffusive processes. So have authors of textbooks on heat transfer (e.g. Kreith, 1973) and micrometeorology (Monteith, 1973). The advantages of such a scheme are many. In heat engineering very complicated systems may be thus represented and heat flows calculated by using Kirchoff's law for electrical circuits. In plant sciences, the transpiring leaf may be represented as a stomatal and boundary layer resistance in series, and hence two useful parameters can be found from measurements of gas exchange (Gaastra, 1959). The definition of the boundary layer, or aerodynamic resistance for heat transfer is:

$$r = \frac{\rho c_p (T_s - T)}{C}, \qquad 2)$$

where ρ is the density of air (kg m^{-2}), c_p is the specific heat of air at constant pressure (J kg^{-1} $°C^{-1}$), and so r has units s m^{-1}. Like h, r depends mainly on the area and shape of the surface and the air speed. The two are simply related:

$$r = \rho c_p / h \tag{3}$$

For mass transfer, relationships corresponding to 2) exist. In general

$$r = (\chi_s - \chi)/F \tag{4}$$

where F is the flux (kg $m^{-2}s^{-1}$), and χ_s and χ are the concentrations of the material in the air at the surface and in the atmosphere, in kg m^{-3}. For the case of water vapour it is usual to work not in concentrations but in vapour pressures. Then, if r is still to be s m^{-1} we need

$$r = \frac{\rho c_p (e_s - e)}{\gamma \;\; F} \tag{5}$$

where γ is the psychrometric constant (kPa $°C^{-1}$).

Increasingly, heat and mass transfers from objects suspended in fluids of any kind are expressed in numbers which have no units, known as non-dimensional numbers. For an introduction to the use of these, the reader is referred to Kreith, 1973, p 315, 385 and Monteith, 1973. The advantage of working with non-dimensional expressions of heat and mass transfer is that results obtained from diverse systems of fluids and surfaces may be more readily applied to a new situation. Thus, from a plethora of experiments already carried out by others it should be possible to calculate the heat transfer to be expected in the case of a lettuce growing behind a shelter belt or a bryophyte cushion immersed in a river.

The non-dimensional number for heat transfer is called the Nusselt number *Nu*, whilst that for mass transfer is the Sherwood number *Sh*. These will be defined fully later, in relation to natural and forced convection, noting here that they do have a simple relation to the boundary layer resistance r

$$Nu = d/r\kappa \tag{6}$$

where d is the dimension of the object parellel to the flow (m) and κ is the thermal diffusivity of the fluid (m^2 s^{-1}). The corresponding dimensionless number for mass transfer is related to r in a similar way

$$Sh = d/rD \tag{7}$$

where D is the diffusion coefficient of the material in whichever fluid is relevant.

NATURAL CONVECTION

Natural convection is a common process of great importance for plants in sheltered places, especially under glass or polythene where protection from the wind is complete. In this mode of convection, a vertical flow develops around the leaf as a result of local warming or cooling of the air in contact with the leaf surface. For instance, if the leaf were in bright sunlight with its stomata shut, its surface temperature might be considerably in excess of the air temperature. Conduction from the surface to the air in contact with the leaf would then cause warming of the air, and thus a decline in the density of this fluid, which would consequently rise to form a vertical plume. If on the other hand the leaf were cooler than the air, as on a windless night under clear skies, the air would fall. In both cases the air moving over the leaf would contribute to heat transfer, as fresh air (unwarmed or uncooled) would be drawn into contact with the surface, thus maintaining a temperature gradient.

The tendency for natural convection to occur depends on a group of five variables, which can be written together in such a way that their units cancel out. Thus, for any given case, we have a dimensionless number which measures the tendency for natural convection to occur. This number is called the Grashof number, *Gr*

$$Gr = fgd^3(T_s - T)/\nu^2 \qquad 8)$$

where f is the coefficient of thermal expansion of air (°C^{-1}), g is the acceleration due to gravity (m s^{-2}), ν is the kinematic viscosity (m^2 s^{-1}) and other variables are as already defined.

The relationship between *Gr* and heat transfer expressed as *Nu* has been investigated many times for various systems. The general form of the relationship, originally guessed from first principles and inferred from dimensional analyses, has been subsequently confirmed experimentally (see Kreith, 1973):

$$Nu = a(Gr\,Pr)^b \qquad 9)$$

where *Pr* is ν/κ, called the Prandtl number, and is 0.70 for air at 20 °C. The values of a and b have been found by experiment. They depend on the geometry of the system, and for most simple shapes like plates, cylinders and spheres, they are well enough established to be listed in books (e.g. Kreith, 1973; Monteith, 1973). Hence, using 8), the value of *Nu* and hence the boundary layer resistance for, say, a flat leaf should be calculable from a knowledge of only two plant attributes: d, the dimension of the leaf, and $(T_s - T)$. The results of such a calculation are presented in Fig. 1.

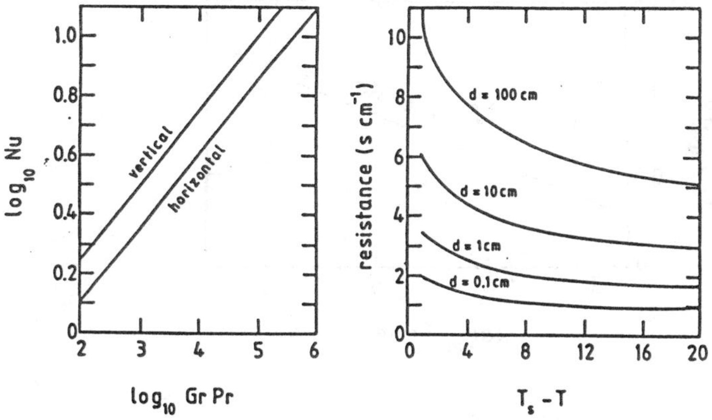

Fig 1. Natural convection relationships calculated from equation 9. On the left hand side is shown the relationship for vertical and horizontal leaves; on the right the aerodynamic resistance is calculated as a function of the leaf-to-air temperature difference, for leaves of various characteristic dimensions.

Recently, this approach was tested over the range of conditions appropriate to leaves in the field (Dixon & Grace, 1983). The test was considered necessary because equation 9 is generally used over a range appropriate to heat transfer problems in engineering. In general, engineers are interested in large and very hot objects, whereas leaves are relatively small and only slightly warmer than the surrounding air. In fact, the limits of applicability of equation 9, defined as a range of *Gr*, are not very well documented. Consulting four text books we found that recommendations were conflicting (see Dixon & Grace, 1983, Table 1).

Our data show that the standard relationship (equation 9, $a = 0.55$, $b = 0.25$) could be relied upon only when *GrPr* exceeded 10^6, whilst below this, in a range frequently occupied by leaves in nature, the heat transfer rate was substantially higher than that calculated, sometimes double (Fig. 2). Similar, very high values of heat transfer at low Grashof numbers have been obtained by other workers (Saunders, 1936; Sinclair, 1970; Kumer & Barthakur, 1971; Schuepp, 1973). When attempting to explain these anomalies, it is perhaps relevant to note that the most serious discrepancies

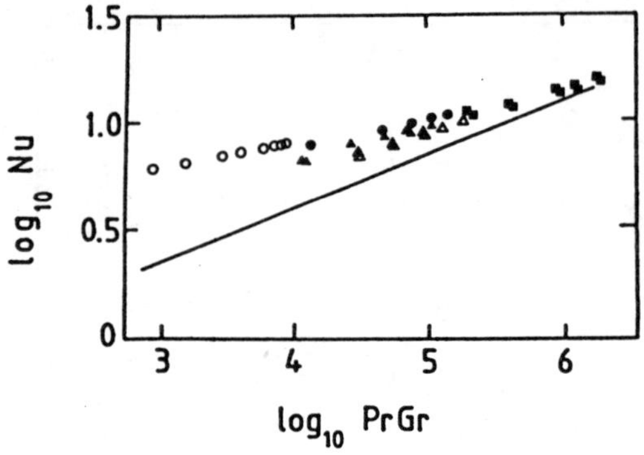

Fig 2. Natural convection relationships calculated from equation 9 (solid line) and found by experiment: O,microphyll, *d*=18 mm; ▲ .Quercus, *d*=46 mm; ●, d=49 mm; △, *Acer*, d=54 mm; ■ *d*=115 mm. From Dixon & Grace (1983).

occur with microphyllous leaves, suggesting that the enhancement of heat transfer is some kind of edge effect. Schuepp (1973) used visualization techniques to demonstrate the whereabouts of convective plumes on a microscale, and showed that the edges of his model leaves were especially active as a source of convective plumes. Another possibility is suggested in Jaluria (1980, p. 265): pure conduction defines the lower limit for *Nu*, and at low *GrPr* this process is substantial in relation to convection. Thus, it seems desirable when working at low values of *GrPr* to replace equation 9 by

$$Nu = m + a(GrPr)^b \qquad 10)$$

where *m* is another coefficient which must be determined experimentally, and which, for our own data (Dixon & Grace, 1983, Fig. 3a) has a value of about 2.

FORCED CONVECTION

In forced convection, heat transfer proceeds at a rate that depends on the wind speed *u* and the dimension of the object *d*, as it is these two parameters which, along with the viscosity of the fluid, determine the average thickness of the boundary layer. Just as the Nusselt number in natural convection can

Table 1. *Relationship between Nu and GrPr for laminar free convection.*

Geometry	Case	Relationship	Range of *Gr Pr*	Definition of *d*
Plate	Vertical	$Nu = 0.55(Gr\,Pr)^{0.25}$	$10^4 - 10^9$	vertical length
	Horizontal hot surface facing up or cool facing down	$Nu = 0.54(Gr\,Pr)^{0.25}$	$10^5 - 10^7$	average chord, except when leaves are long and thin, when width is best.
	Horizontal hot surface down or cool surface up	$Nu = 0.27(Gr\,Pr)^{0.25}$	$3 \times 10^5 - 3 \times 10^{10}$	
Cylinder	Horizontal	$Nu = 0.53(Gr\,Pr)^{0.25}$	$10^4 - 10^9$	diameter
	Vertical	$Nu = 0.55(Gr\,Pr)^{0.25}$	$10^4 - 10^9$	diameter
Sphere		$Nu = 2 + 0.43(Gr\,Pr)^{0.25}$	$1 - 10^5$	diameter

be correlated with the Grashof number, so in forced convection the Nusselt number is correlated with the Reynolds number *Re*, defined as ud/v. For flat plates:

$$Nu = 0.66\,Re^{0.5}Pr^{0.33} \qquad 11)$$

this applies to systems involving laminar flow. In turbulent flows the relationship displays a nearly-linear relationship with windspeed

$$Nu = 0.66Re^{0.8}Pr^{0.33} \qquad 12)$$

There is a question here as to whether the boundary layer over a leaf is laminar or turbulent. In classical studies in fluid dynamics, where a smooth flat plate was the usual test object and the airflow was generally laminar, a transition from laminar to turbulent flow was observed at around $Re = 10^5$. This critical point was always found to be sensitive to the introduction of turbulence into the wind tunnel or to any roughness on the plate. Studies of leaves in laminar or turbulent flows show that these objects are aerodynamically much rougher than smooth flat plates: there is some degree of turbulence in the boundary layer in virtually all conditions (Grace & Wilson, 1976).

This tendency for turbulence to occur over the leaf surface even at low values of *Re* may explain why many experimental data exceed the values calculated from 11). For a review of such data the reader is referred to

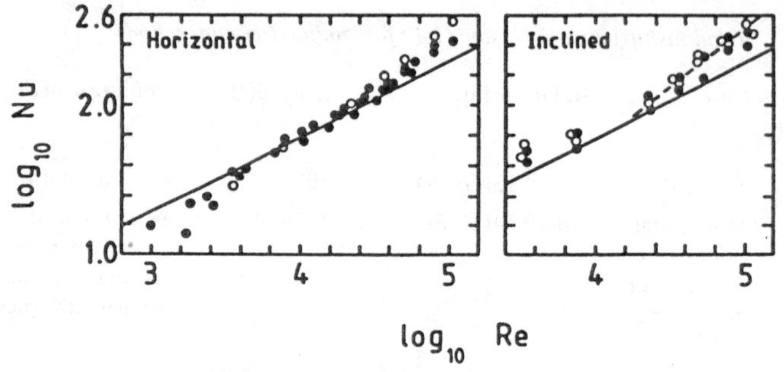

Fig 3. Forced convection relationships calculated from equation 11. The data points are from wind tunnel experiments on large metallic leaves held horizontal or inclined at 10° and 45° in a turbulent air flow. The broken lines represent the corresponding relationship for a turbulent boundary layer. For details see Grace *et al* (1980).

Grace, 1981 p. 34. Another possible reason for any discrepancy may simply be poor technique, especially when using water vapour as the diffusing species to measure mass transfer (Grace & Wilson, 1976). A much more reproducible technique, which is almost certainly more accurate in an absolute sense as well, involves the careful recording of the cooling rate of model leaves which have been heated to a few degrees above ambient (Grace *et al*, 1981). In an early series of experiments on large-leaved species we found good agreement between observed and calculated heat transfer (Grace *et al*, 1981). There were indications that leaf minutiae could influence the result, especially at high values of *Re* ($>10^4$). On the whole though, this series of experiments provides some grounds for confidence in the existing relationship (equation 11), despite the demonstration that the boundary layer is often turbulent to some degree (Fig. 3). The equation for heat transfer across a turbulent boundary layer (equation 12) should not be used in any case at less than $Re = 17000$ as it was not developed for this purpose and in fact, yields lower values of *Nu* than does equation 11). Boundary layer resistances calculated from equation 11 are plotted to illustrate the effect of leaf size and wind speed (Fig. 4).

Recently, other experiments have been carried out in our laboratory to assess more carefully the effect of turbulence in the air impinging on the leaf, and the effect of various designs of leaf (Dixon, 1982). These experiments will be described in full elsewhere and only summarised in the

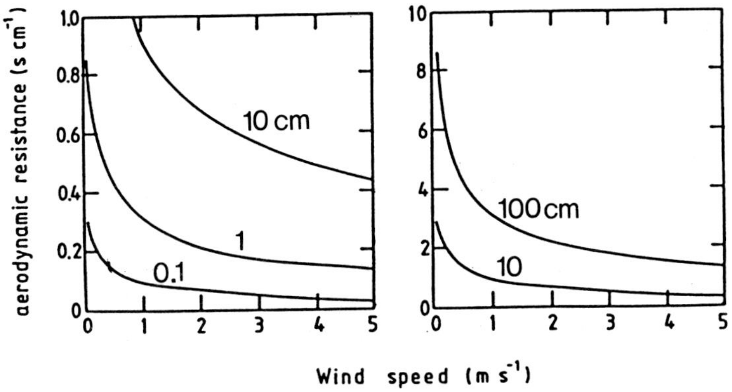

Fig 4. Forced convection. The aerodynamic resistance is shown as a function of wind speed for various characteristic dimensions.

present chapter. The main conclusion was that turbulence in the free stream, and possession of serrated or pleated margins by the leaf, did indeed increase the rate of heat transfer, but by not more than 20%. Previously, other workers have obtained as much as double the calculated rates and attributed these high rates to turbulence (see reviews in Grace, 1977, 1981). It now seems unlikely that these extremely high rates were caused by anything other than experimental error.

UNRESOLVED PROBLEMS IN HEAT TRANSFER FROM LEAVES

1. *Mixed convection*

There must be many cases in nature where natural and forced convection both contribute substantially to the total convective heat transfer. It is not at all obvious how such cases can be dealt with. In an extreme situation, which might occur in a growth cabinet, we can imagine a vertical flow of air upwards, caused by bouyancy to be either augmented, if the flow of air in the cabinet is upwards, or diminished if the flow of air in the cabinet is downwards. Thus in the first case the two processes would be additive, whilst in the second they would tend to cancel each other out. For the most common case, that of an upward vertical plume caused by bouyancy, and a horizontal wind, the net effect would presumably lie between these two.

Kreith (1973) shows that the ratio Gr/Re^2 gives a qualitative indication of the influence of buoyancy on forced convection. When the Grashof number is the same order of magnitude or larger than the square of the

Reynolds number, free convection effects 'cannot be ignored'. Examples of full analyses for individual cases are given in Jaluria (1980), but in the absence of any theoretical or experimenatal work on plant leaves it is not clear how to proceed. It is perhaps wisest to calculate both natural and forced convection rates. If they are of comparable magnitude the two values should either be added, or the larger taken, depending on whether the two flows are in the same direction or not.

2. *Leaf angle*

Several studies have shown that the influence of leaf angle on heat transfer by forced convection is rather small (Thom, 1968; Parkhurst *et al*, 1968; Grace *et al*, 1981). In a most thorough study, Parkhurst *et al* (1968) showed that the influence of leaf angle became great only in long thin leaves, when air flow was along the longest dimension. Much insight into local variation in r over the surface of an inclined leaf was obtained in a mass transfer experiment by Chamberlain (1974).

In natural convection, there are substantial and well-established differences between heat transfer from horizontal and vertical surfaces, and smaller differences between upward and downward facing horizontal surfaces. Appropriate coefficients to use in equation 9) when dealing with such cases are tabulated by Kreith (1973). The case of flat surfaces inclined at an angle is discussed on pages 396-7 in Kreith (1973).

The case of leaf flutter is interesting as it involves variations in leaf angle coupled with variations in flow. Flutter is a periodic phenomenon, involving cycles of lift and stall. Unlike the cases usually considered, it is not a steady state situation but one which cycles at a period of 1-50 Hz. Local heat transfer from a small patch on a fluttering leaf can be measured with a glue-on hot-film anemometer. The results of one study suggested that flutter simply introduces sinosoidal variations around the mean value which hardly differs from that obtained if the leaf is held in one position (Grace, 1978). From this, and several other experiments (Raschke, 1956; Parkhurst *et al*, 1968; Parlange *et al*, 1971) it appears that fluttering does not have a very large effect in relation to heat transfer from a stationary leaf at the same windspeed.

3. *Heterogeneity of leaf surfaces*

The leaves of some species are peculiarly folded or pleated. A glance at such a leaf surface in sunny weather reveals a pattern of sunlit and shaded areas. It is doubtful whether the input of solar radiation to such a leaf can be estimated with much accuracy; the spatial variation over the leaf surface might vary by at least an order of magnitude. As far as heat dissipation is concerned, serrated or lobed margins and prominent veins will add further to the heterogeneity in surface temperature as the air flow is likely to be complex (see Grace, 1983, p. 40). Some idea of the variation that can occur in surface temperature is seen on infra red images of leaf surfaces (Clarke & Wigley, 1974). The only way to find the convective heat transfer from such subjects may be to resort to a direct measurement technique rather than attempting to apply standard formulae. It is noteworthy that most of these remarks apply equally to animal surfaces, a point to be returned to in a later chapter by Bell.

4. *Finding d, the characteristic dimension*

The characteristic dimension is the effective length of surface over which the boundary layer grows. For a flat rectangular plate it is simply the length parallel to the flow. For other standard shapes the appropriate dimension of the leaf can be looked up (e.g. Monteith, 1973, p. 224-5). For simple leaves, the procedure outlined by Monteith, that of taking the mean of chords parellel to the flow appears adequate on both intuitive and experimental grounds (Parkhurst *et al*, 1968). Compound leaves pose special problems. If the lobes are very close together it may be that the thickness of the boundary layer remains constant as the air traverses the gap between lobes, and then continues to grow again when the new surface is reached. If on the other hand the lobes are distant there is likely to be complete mixing of the air in such a way that the boundary layer must start to grow all over again. Experimental data bearing on this problem are given by Parkhurst *et* al, (1968); Vogel, (1970) and Grace *et al*, (1981). Similar questions are raised when attempting calculation of heat or mass flow from groups of leaves. Landsberg & Thom (1971) and Landsberg & Powell (1973) attempt to devise an empirical shelter factor. Alternative approaches are illustrated in Kreith (1973) and by Wilson & Crowther in a later chapter.

MASS TRANSFER

By mass transfer in this context is meant the transfers of gaseous materials, especially H_2O, CO_2 and pollutant gases, which occur by diffusion and turbulent diffusion across the boundary layer of the leaf. Very small particles may behave like gases in this respect, but larger ones do not. For a discussion of particle transport the reader should see Chamberlain & Little (1981).

The mechanism of mass transfer is analogous to that of heat transfer by conduction or convection. Both entities are transferred in a direction which tends to reduce an existing gradient. Mass transfer by molecular diffusion is analogous to heat transfer by thermal agitation in still air. In moving air, both entities are caried along in the streamlines or in the eddies that constitute turbulent flow. Consequently, both are influenced by the fluid dynamic properties of the system. Thus, heat transfer rates and mass transfer rates through boundary layers should be simply related.

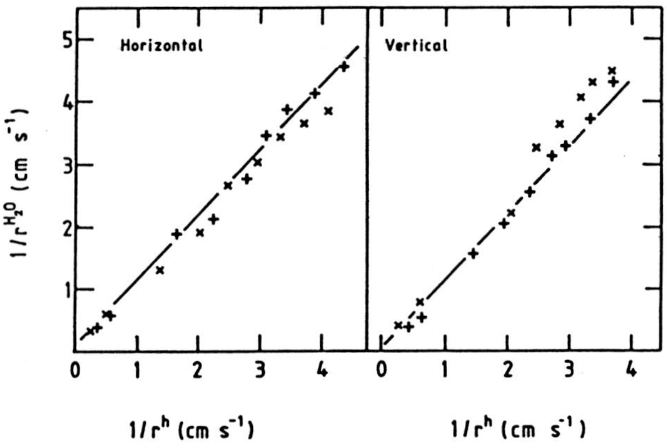

Fig 5. Relationship between aerodynamic resistances for heat and water vapour transfer. Determined using models of *Populus alba* leaves in a horizontal or vertical posture over the range 0-5.2 m s^{-1} in laminar (+) or turbulent (×) flow. Grace & Slack (unpublished).

Recalling the Nusselt number for heat transfer across a laminar boundary layer

$$Nu = 0.66\, Re^{0.5} Pr^{0.33}$$

The analogous expression can be written for mass transfer and the coefficients found by experiment:

$$Sh = 0.66\, Re^{0.5} Sc^{0.33} \qquad 13)$$

where Sc is the Schmidt number, given by ν/D. Recalling that Pr is ν/κ the relationship between Sh and Nu must be

$$Sh = Nu(\kappa/D)^{0.33} \qquad 14)$$

whilst the ratio of diffusive resistances can be found by inserting an appropriate value for the diffusion coefficient, D. For water vapour in air

$$\frac{r^{H_2O}}{r^{h}} = \left[\frac{\kappa}{D^{H_2O}}\right]^{0.66} = 0.93 \qquad 15)$$

Note that this relationship should still hold for a turbulent boundary layer (using equation 12 in the same way as we have just used equation 11).

For the case of *Populus alba*, an attempt was recently made to verify this relationship (Slack & Grace, unpublished). Heat transfer was measured by the method already referred to, whilst evaporation was measured by the filter paper technique described by Jarvis (1971), but using four thermocouples on the surface to guarantee a good determination of surface temperature, necessary in the calculation of r^{H_2O}. The result showed good agreement with the factor of 0.93 derived above (Fig. 5). Earlier work by Thom (1968) using bromobenzene, methyl salicylate and water as diffusing species suggests that such calculations can be applied to any gas as long as the diffusion coefficient has been established.

For the case of natural convection, arguments of similarity equally apply, as the convective currents that are set up will cause transport of mass as well as heat. A similar approach for laminar convection gives

$$Sh = Nu(\kappa/D)^{0.25}$$

so that

$$\frac{r^{H_2O}}{r^{h}} = \left[\frac{\kappa}{D^{H_2O}}\right]^{0.75} = 0.915$$

Fortunately, the difference between this and the result for forced convection is small enough to be ignored.

FINAL REMARKS

The prospect of estimating rates of heat and mass transfers to and from leaves from a knowledge of the appropriate parameters of the physical environment and the size of the leaves is probably better for some species than others. For those plants with simple leaves on long petioles there should be no great difficulty. For plants with crowded leaves, or leaves pressed to the stem, there are considerable difficulties, at least until more work has been done on the problem of defining an appropriate characteristic dimension.

It should be borne in mind that the operative work in the previous paragraph is 'estimating'. An engineer designing a power station would build in a substantial margin of safety, in the knowledge that there will be circumstances in which the estimates will be wrong. A biologist cannot expect to do any better than this, and may do less well, for the system under study is generally less tidy.

To predict surface temperatures and transpiration rates, once the boundary layer resistance has been estimated, requires other information. Specifically, the stomatal resistance, air humidity and net radiation absorbed by the leaf is required. Thereafter, transpiration rates may be found by applying the Penman-Monteith equation (Dixon & Grace, 1984). Alternatively the energy balance may be solved by an iterative procedure (Grace, 1983) to obtain surface temperature, convective heat transfer and rate of transpiration.

REFERENCES

Chamberlain, A.C. (1974). Mass transfer to bean leaves. Boundary-layer Meteorology, **6**: 477-486.

Chamberlain, A.C. & Little, P. (1981). Transport and capture of particles by vegetation. In Plants and Their Atmospheric Environment. J. Grace, E.D. Ford & P.G. Jarvis (Eds.) pp 147- Oxford: Blackwell Scientific Publications.

Clark, J.A. & Wigley, G. (1974). Heat and mass transfer from real and model leaves. In Heat and Mass Transfer in the Biosphere 1. Transfer Processes in the Plant Environment. pp 353-365. New York: Wiley.

Dixon, M. (1982). Effect of Wind on the Transpiration of Young Trees. Ph.D. Thesis, University of Edinburgh.

Dixon, M. & Grace, J. (1983). Natural convection from leaves at realistic Grashof numbers. Plant, Cell & Environment. **6**: 665-670

Dixon, M. & Grace, J. (1984). Effect of wind on the transpiration of young trees. Annals of Botany (in press).

Gaastra, P. (1959). Photosynthesis of crop plants as influenced by light, carbon dioxide, temperature and stomatal diffusion resistance. Mededelingen Landbouwhogesch Wageningen, **59**: 1-68.

Grace, J. (1977). Plant Response to Wind. London: Academic Press.

Grace, J. (1978). The turbulent boundary layer over a flapping *Populus* leaf. Plant, Cell & Environment, **1**, 35-38.

Grace, J. (1981). Some effects of wind on plants. In Plants and Their Atmospheric Environment, British Ecological Society Symposium 21. J. Grace, E.D. Ford & P.G. Jarvis (Eds.) pp 31-56. Oxford: Blackwell Scientific Publications.

Grace, J. (1983). Plant-atmosphere Relationships. London: Chapman & Hall.

Grace, J., Fasehun, F.E. & Dixon, M. (1980). Boundary layer conductance of some tropical timber trees. Plant, Cell & Environment, **3**: 443-450.

Grace, J. & Wilson, J. (1976). The boundary layer over a *Populus* leaf. Journal of Experimental Botany, **27**: 231-241.

Jaluria, Y. (1980). Natural Convection Heat and Mass Transfer. Oxford: Pergamon Press.

Jarvis, P.G. (1971). The estimation of resistances to carbon dioxide transfer. In Plant Photosynthetic Production, Manual of Methods. Z. Sestak, J. Catsky & P.G. Jarvis (Eds.) pp 556-631. The Hague: Junk.

Kowalski, G.J. & Mitchell, J.W. (1975). Heat transfer from spheres in the naturally turbulent out-door environment. Presented at the American Society of Mechanical Engineers Winter Annual Meeting 1975. Paper no. 75-WA/HT-57.

Kreith, F. (1973). Principles of Heat Transfer. Third Edition. New York: Harper & Row.

Kumer, A. & Barthakur, N. (1971). Convective heat transfer measurements on plants on a wind tunnel. Boundary-layer Meteorology, **3**: 454-467.

Landsberg, J.J. & Powell, D.B.B. (1973). Surface exchange characteristics of leaves subject to mutual interference. Agricultural Meteorology, **12**: 169-184

Landsberg, J.J. & Thom, A.S. (1971). Aerodynamic properties of a plant of complex structure. Quarterly Journal of the Royal Meteorological Society, **97**: 565-570.

Linacre, E.T. (1974). Determinations of heat transfer coefficients of a leaf. Plant Physiology, **39**: 687-690.

Mitchell, J.W. (1976). Heat transfer from spheres and other animal forms. Biophysical Journal, **16**: 561-569.

Monteith, J.L. (1973). Principles of Environmental Physics. London: Edward Arnold.

Murphy, C.E. & Knoerr, K.R. (1977). Simultaneous determinations of the sensible and latent heat transfer coefficients for tree leaves. Boundary-layer Meteorology, **11**: 223-241.

Parkhurst, D.F., Duncan, P.R., Gates, D.M. & Kreith, F. (1968). Wind tunnel modelling of convection of heat between air and broad leaves of plants. Agricultural Meteorology, **5**: 33-47.

Parlange, J., Waggoner, P.E. & Heichel, G.H. (1971). Boundary layer resistance and temperature distribution on still and flapping leaves. Plant Physiology, **48**: 437-442.

Raschke, K. (1956). Über die physikalischen Beziehungen zwischen Wärmeübergangszahl, Strahlungsaustausch, Temperatur und Transpiration eines Blattes. Planta, **48**: 200-238.

Saunders, O.A. (1936). The effect of pressure upon natural convection in air. Proceedings of the Royal Society, **157a**: 278-291.

Schuepp, P.H. (1973). Model experiments on free convection heat and mass transfer of leaves and plant elements. Boundary-layer Meteorology, **3**: 454-467.

Sinclair, R. (1970). Convective heat transfer from narrow leaves. Australian Journal of Biological Science, **23**: 309-321.

Thom, A.S. (1968). The exchange of momentum, mass and heat between an artificial leaf and the airflow in a wind tunnel. Quarterly Journal of the Royal Meteorological Society, **94**: 44-55.

Thorpe, M.R. & Butler, D.R. (1977). Heat transfer coefficients for leaves on orchard apple trees. Boundary-layer Meteorology, **12**: 61-73.

Vogel, S. (1970). Convective cooling at low airspeeds and the shapes of broad leaves. Journal of Experimental Botany, **21**: 91-101.

Progress in Biometeorology, 1985, Vol. 2, pp. 17-36

FLOW VISUALISATION AND THE STUDY OF SHELTER EFFECTS FOR VEGETATION AT THE MICROSCALE

C.E. Wilson and J.M. Crowther
Department of Applied Physics,
University of Strathclyde,
107 Rottenrow,
Glasgow, G4 0NG.

INTRODUCTION

In almost all plants the leaves are grouped together in bunches around a stem. Attempts to calculate heat, mass and momentum transfer between leaves and the atmosphere, or to calculate leaf temperatures from a knowledge of environmental parameters, ought not to assume that the leaves are aerodynamically independent. To a greater or lesser extent they shelter one another.

The aim of this paper is to increase our understanding of how shelter between coniferous needles occurs. This is done by investigating the pattern of flow using visualisation techniques and aims at constructing a model for the mechanism of shelter within individual shoots under simplified conditions. This model for shelter can then be tested by comparing values of heat and momentum transfer calculated theoretically for the case of the unsheltered shoot and for a shoot sheltered according to the flow model. This would give a theoretical "shelter factor" which could be compared with the experimental values for shelter found by Landsberg & Thom (1971).

In the literature, the term shelter is used in different ways. Often it is a modifying factor to relate theoretical values for transfers to experimental data. In fact Shuttleworth (1976) describes shelter factors as "at worst, no more than a mathematical device in which to shroud the complexity of elemental interference introducing apparent simplicity into the equations." As such, shelter factors have been incorporated into several mathematical models of the vegetation-atmosphere interaction. However, within this paper the shelter factor p is defined in terms of Landsberg & Thom (1971) where

$$p = \frac{C'}{C} \tag{1}$$

that is the ratio of the theoretically calculated value of a transfer coefficient to the experimentally found result. In calculating C' it is assumed all elements are to be independent and perpendicular to the flow. For the cases of mathematical models of a crop canopy it is only necessary to know a numerical value for shelter as a mean for the canopy or a given level within the canopy. However a greater understanding of the mechanisms involved has interesting implications for plant physiology and meteorology at the microscale.

In Landsberg & Thom (1971) shelter was defined as "mutual aerodynamic interference", implying that it is changes in the flow pattern due to the presence of the foliage that give rise to the shelter effect. Other writers such as Shuttleworth (1976) suggest that microscale changes occur also in temperature and vapour pressure but it is thought that these would have little significance compared with the changes in wind speed and direction.

It was therefore decided to investigate the flow modifications using flow visualisation techniques.

EXPERIMENTAL TECHNIQUE

The essence of the experiment is to examine the modified flow around a shoot in a wind tunnel by making it visible and photographing it. These data can then be described in quantitative terms by fitting the flow patterns to plots of calculations of potential flow, relying on the fact that upstream of the separation point, even at high Reynolds numbers, potential flow is a good approximation to reality.

The technique of flow visualisation was chosen because the measurement of a fluctuating flow around a body of complex geometry is extremely difficult with conventional pressure or windspeed measuring techniques. The alternative of flow visualisation gives a relatively quick route to discovering and recording the modified pattern of flow due to aerodynamic interference.

The wind-tunnel used for this study is an open duct, 4 m in length with the cross-section of the working section being 0.3 m by 0.3 m. Plate glass panels are set into the top and sides of the working-section to improve the optics for photography. The fan speed can be varied and with honeycomb screens in position the flow is predominately laminar or near laminar. The wind speed was measured using a Pitot tube and a micromanometer.

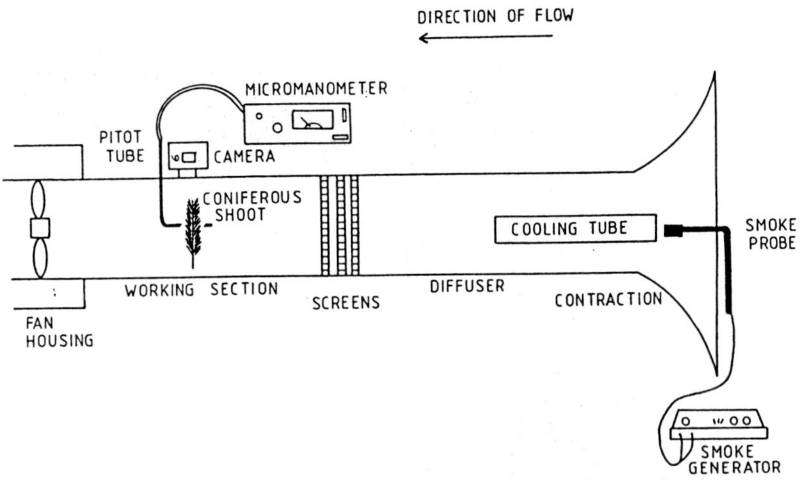

Fig 1. Wind Tunnel with Flow Visualisation Apparatus.

To render the flow visible a tracer is introduced which follows the direction of flow closely. In this case a cooled smoke was used. It was conveniently broken into individual "sources" by the honeycomb screens to give discrete streamlines. The flow was photographed from above and illuminated from both sides (Figures 1 & 2). The best photographs were obtained by using a thin horizontal plane of light through the working section. This thin plane of light was provided by two slide projectors giving a focused parallel beam. The width of the beam was determined by razor blades mounted in standard slide frames. To increase contrast and reduce reflections the wind-tunnel was blacked out from the inside. Photographs were taken using a fast film.

Description of Flow

The photographs (Figure 3) show the flow to be not so much "mutual aerodynamic interference" at the intra-shoot level as collective or

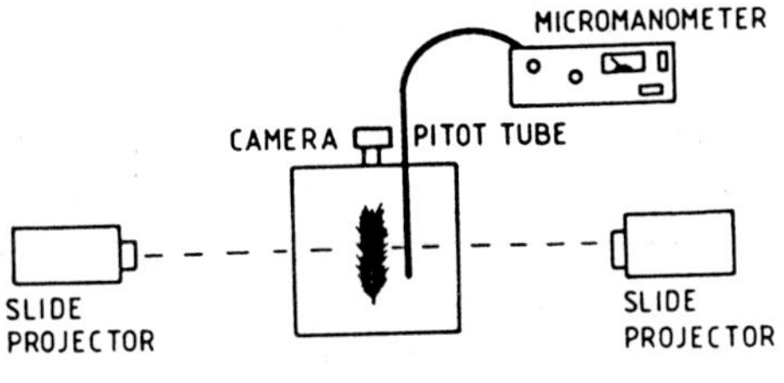

Fig 2. Cross-section of Wind Tunnel and Flow Visualisation Apparatus.

Fig 3. Photograph of Flow Around a Shoot at 1 m s^{-1}.

composite aerodynamic interference. With the Reynolds number of an average needle being 80 and of a twig being 270 it is unlikely that either of these elements would produce such a wide wake or disruption of flow. Instead it appears that the needles act collectively as a porous bluff body. The upstream diversion of flow, the separation streamline and the wide wake are indicative of a bluff body much larger than any individual part of the shoot. This gave reason to suspect that the increasing density of foliage near the twig was causing a collective effect, giving a result which could be described in terms of a cylinder with an effective aerodynamic diameter.

In order to quantify this effective aerodynamic diameter the flow lines on the photograph were matched with plots of two-dimensional potential flow around cylinders with diameters D_c greater than the twig but smaller than the total diameter of the shoot. These plots were constructed by calculating the solutions to the potential equations for velocity u, and velocity potential ϕ:

$$u = -\nabla\phi, \qquad \nabla^2\phi = 0 \qquad (2)$$

with boundary conditions:

$$u \to u_0 \quad \text{as} \quad r \to \infty$$
$$u_r = (u.r) \to 0 \quad \text{as} \quad r \to D/2$$

The values of the velocity components were calculated at short intervals along the x-axis and the streamlines plotted (c.f. Equation (5)). An example of such a plot is given in Figure 4. These plots were photocopied onto transparencies to be matched with flow lines on the photographs upstream of the separation point. Using this method of determining the effective aerodynamic diameters the following results were obtained, accurate to ± 0.5 mm.

Table 1 *Variation of Effective Aerodynamic Diameter with Windspeed.*

Experiment Number	Windspeed ($m\,s^{-1}$)	Effective Aerodynamic diameter(mm)	Shoot Density $\sigma = (A_N + A_T)/A_S$ (Landsberg & Thom 1971)
1	0.40	16	2.0
2	0.44	16	2.0
3	0.82	14	2.0
4	1.00	13	1.8

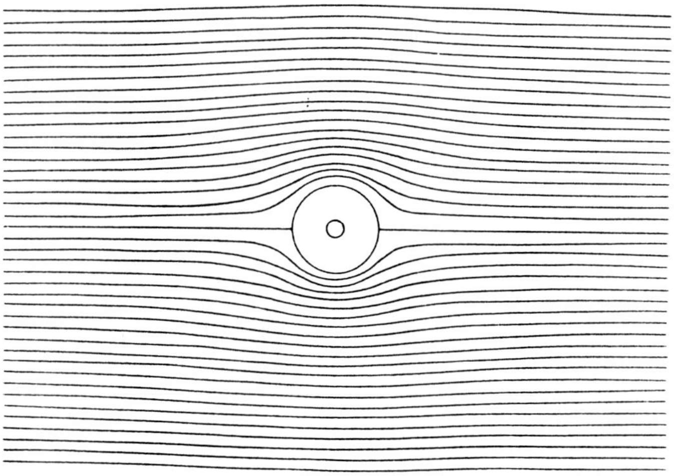

Fig 4. An Example of a Plot of Potential Flow Around a Cylinder.

As one would intuitively expect, the effective aerodynamic diameter decreases as the windspeed increases and the boundary layer becomes thinner. In order to test this modified flow pattern model, shelter factors were calculated using this method and compared with those found experimentally by Landsberg & Thom (1971).

CALCULATION OF THEORETICAL SHELTER FACTORS

In this section an attempt will be made to calculate shelter factors under certain simplifying assumptions:

1. that transfer to and from the twig will occur as if the twig had a radius equal to the effective aerodynamic radius;
2. that the needles will be exposed to the potential flow associated with the windspeed and the effective aerodynamic radius, and that no transfer will occur for the portion of the needle within the aerodynamic diameter of the shoot;
3. that the needles will be considered as cylinders which have the same surface areas,

$$d \simeq \sqrt{2(a^2 + b^2)}$$

4. that the shoot has an average axial symmetry and is as shown in Figure 5;
5. that end effects may be neglected for both twig and needles.

The other data required for the model are based on average measurements for five shoots and are given in Table 2. A more detailed analysis of the method for calculation is given in the Appendix.

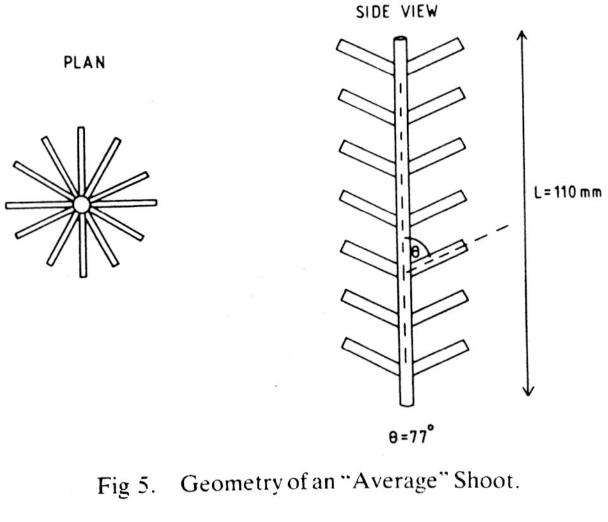

Fig 5. Geometry of an "Average" Shoot.

Calculation of Drag and Heat Transfer for the Unsheltered shoot

In this calculation the method is based on that of Landsberg & Thom (1971) for the case of an unsheltered shoot. All the elements are considered as independent and perpendicular to the flow, even though it is clearly unrealistic to do so.

The drag is calculated as a summation over all elements, needles and twigs, as described in the Appendix:

$$F_a = \tfrac{1}{2}\rho u_0^2 \Sigma_i A_i C_{Di} \qquad (3)$$

In the calculation it is assumed that $u_0 = 1.0$ m s^{-1}, and for the needles $A = 1.714 \times 10^{-5}$ m^2, $Re = 68$, $C_D = 1.915$; for the twigs $A = 4.40 \times 10^{-4}$ m^2, $Re = 267$, $C_D = 1.361$. The density of air at 20° C is 1.2 kg m^{-3} and so

$$F_a = 4.24 \text{ mN}$$

The heat transfer is calculated as follows

$$Q_a = k(T_S - T_A)\, \Sigma_i \left[\frac{NuS}{d}\right] \qquad (4)$$

where S is the surface area, d the diameter and Nu the Nusselt number. $S/d = \pi l$ for needles and πL for the twig (ignoring end effects).

For the needles $Nu = 4.60$ and for the twigs $Nu = 8.72$. For air at 20° C, $k = 25.7$ mW m^{-1} K^{-1}, and assuming $T_S - T_A = 1$K, we have

$$Q_a = 1.31 \text{ W}$$

Calculation of Drag for the Sheltered Case

Here the assumptions are as stated at the beginning of this section, and take account of both angle of incidence and modifications of flow velocity.

Let us first calculate the drag for the twig using the effective aerodynamic

Table 2 *Model Parameters Based on an Average of Five Shoots.*

Item	Symbol	Value
No. of needles	n	197
Needle length(mm)	l	16.8
Needle major axis(mm)	a	1.22
Needle minor axis(mm)	b	0.76
Needle effective diameter (mm)	d	1.02
Needle angle(°)	θ	77
Twig length(mm)	L	110
Twig diameter(mm)	D	4.0

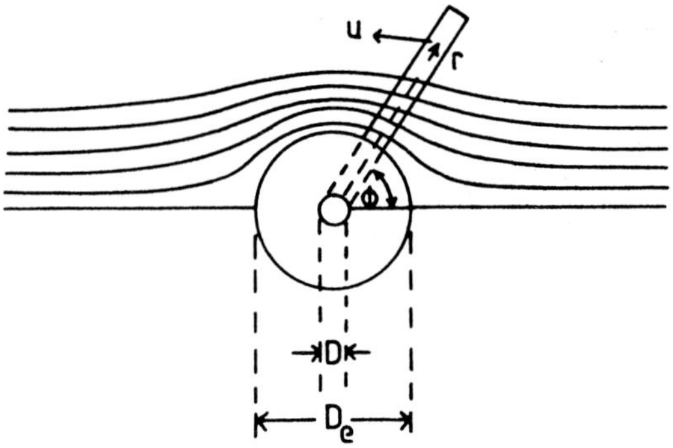

Fig 6. Components of Flow.

diameters; in this case $D_e = 13.0$ mm, Reynolds number $Re = 867$ and drag coefficient $C_D = 1.014$ giving a drag force 8.70×10^{-4} N at 1 m s^{-1}.

For the needles the situation is more complicated since the local windspeed and direction relative to the needle are varying. It is therefore necessary to perform an integration as follows.

Figure 6 shows a single needle exposed to potential flow. The integration must be performed from the intersection of the needle with a cylinder of effective diameter D_e out to the end of the needle. If the cylindrical polar coordinates are denoted by (r,ϕ,z) then the radial and tangential velocity components are

$$u_r = -u_o \cos\phi(1 - D_e^2/4r^2) \quad (5)$$

$$u_\phi = u_o \sin\phi(1 - D_e^2/4r^2) \quad (6)$$

The velocity parallel to the needle is (c.f. Figure 7)

$$u_p = u_r \sin\theta \quad (7)$$

and the velocity perpendicular to it is

$$u_n = (u_r^2 \cos^2\theta + u_\phi^2)^{1/2} \quad (8)$$

For the calculations of drag, only u_n will be considered as effective, but the drag force will be resolved in the direction of the undisturbed flow.

The drag force on an infinitesimal element is

$$dF(\phi) = \tfrac{1}{2}\rho u_n \frac{d}{\sin\theta} C_D\left(\frac{u_n d}{\nu}\right)(u_n)_o dr \quad (9)$$

where $(u_n)_o$ is the component of u_n in the direction $\phi = 180°$. The radial integration gives

$$F(\phi) = \int_{D_e/2}^{D_e/2 + l\sin\theta} dF(\phi) \qquad (10)$$

The integration over all needle angles is performed as follows

$$F_b = \frac{n}{2\pi} \int_0^{2\pi} F(\phi)d\phi + 8.70 \times 10^{-4}\mathrm{N} \qquad (11)$$

The calculated value is

$$F_b = 3.05\ \mathrm{mN}$$

If, however it is argued that the wake effect is included in the effective aerodynamic diameter of the shoot then it is a better approximation to omit the contribution of needles situated in the wake. This is effected by noting that the wake is enclosed by $\phi = 142.5°$ to $\phi = 217.5°$. Applying this in the above integral gives a value

$$F'_b = 2.96\ \mathrm{mN}$$

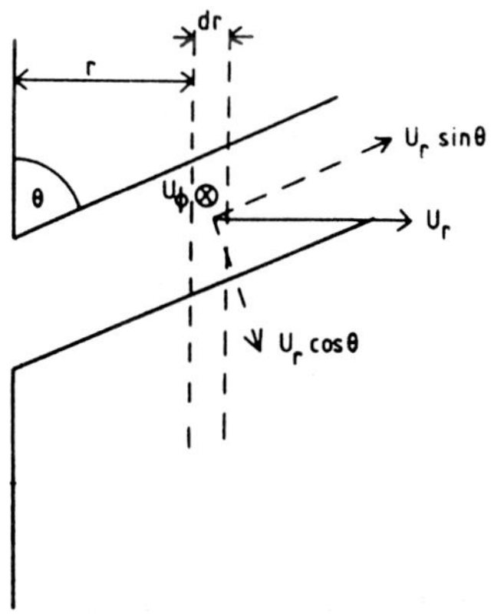

Fig 7. Resolution of Flow into Parallel and Perpendicular Components from Radial and Tangential Components.

Calculation of Heat Transfer

This proceeds in a similar way. As explained in the Appendix, for the needles

$$Nu = 0.39 + 0.51 d^{1/2} v^{1/2} (u_n^2 + 0.04 u_p^2)^{1/4} \quad (12)$$

and for the twig

$$Nu = 0.39 + 0.51 (867)^{1/2}$$

Using these formulae in the calculation

$$Q_b = 0.985 \text{ W}$$

Similarly, omitting needles in the wake gives

$$Q'_b = 0.959 \text{ W}$$

These results give shelter factors that are summarised in Table 3 together with data from Landsberg & Thom (1971).

Table 3 *Shelter Factors*

	Drag	Heat
Landsberg & Thom (1971) (experimental) at 3 m s^{-1}	1.91	1.91
Calculated including wake at 1 m s^{-1}	1.39	1.36
Calculated excluding wake at 1 m s^{-1}	1.43	1.52

The shelter factor of 1.91 attributed to Landsberg & Thom (1971) was estimated for the model (Figure 5) using the shoot density of 0.94 calculated for the shoot outline (i.e. consistent with their method) and their Equation 12.

The experimental results and theoretical results in Table 3 are not strictly comparable because the windspeed is different and because the former used a compound shoot as opposed to a single shoot. Nevertheless, the more realistic calculations excluding the wake comes within approximately 20% of the experimental value: a difference which could be explained by interference between elements of the *compound* shoot.

CONCLUSIONS

The flow visualisation photographs show a considerable modification of the flow by the needles and twig as a porous bluff body. It is possible to study in detail the effect of changing windspeed on the flow and to estimate transfer properties like drag and heat loss. Contrary to the findings of Landsberg & Thom (1971), there will be significant differences between the shelter factors for heat and momentum transfer, and the shelter factors will also vary with windspeed.

Given the obvious importance of small scale changes in both wind speed and direction for transfer processes, flow visualisation offers a valuable tool in micrometeorological studies.

APPENDIX

Analysis of Drag

The drag force on an object is the viscous force which opposes the relative motion of the object and the fluid in which it is immersed. It is usual to express the drag force F, in terms of a dimensionless drag coefficient C_D:

$$F = \tfrac{1}{2}\rho u^2 A C_D \qquad (A1)$$

where ρ is the density of the fluid and A is the projected area of the object in the direction of flow (or relative motion).

Figure A1 shows the form of the drag coefficient C_D as a function of Reynolds Number Re for a cylinder at normal incidence.

In calculating the drag for coniferous foliage we may consider, as a first approximation, that twigs and needles are cylindrical elements. For Sitka spruce a typical diameter of a twig would be 4 mm, but for the needle the

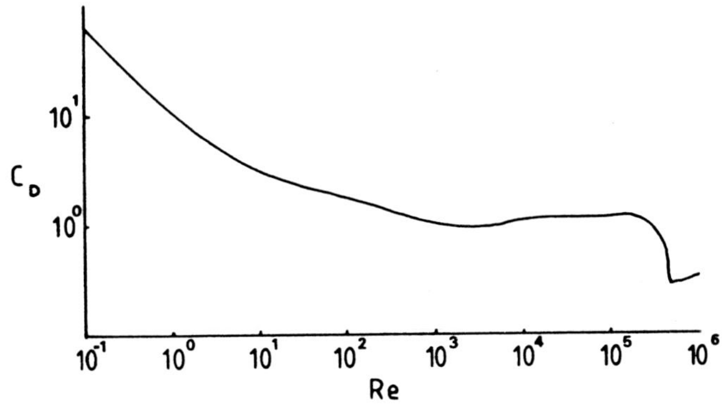

Fig A1. Drag Co-efficient as a Function of Reynolds Number. (from Schlichting 1968, p17.).

shape is more elliptical than cylindrical with major and minor axis of around 1.2 mm and 0.75 mm respectively. Taking a windspeed of around 1 m s^{-1} and a value of kinematic viscosity $\nu = 0.15 \times 10^{-4}$ m^2 s^{-1} (corresponding to air at 20° C) gives the values of Reynolds Number and drag coefficients listed in Table A1.

Table A1 *Representative Drag Co-efficients for Twigs and Needles.*

Element	u	d	Re	C_D
Twig	1.0	4.00	267	1.3
Needle	1.0	0.75	50	2.0
Needle	1.0	1.20	80	1.8

The data in Figure A1 will be represented by the following expressions (accurate to better than 5%):

$$\begin{aligned} &Re < 4 && C_D = 10\,Re^{-0.77} \\ &4 < Re < 1000 && C_D = 5.5Re^{-0.28} \\ &1000 < Re < 6000 && C_D = 1.0 \end{aligned} \tag{A2}$$

The data in Figure A1 are taken for quasi-infinite cylinders, but for finite cylinders, the drag coefficient may be considered to depend on the ratio of length to diameter (l/d) for the cylinder. Rouse (1978) gives the following data for $Re = 10^5$.

Table A2 *Effect of l/d on Drag of Cylinders Normal to Flow at High Reynolds Numbers*

l/d	1	5	20	∞
C_D	0.63	0.74	0.90	1.20

The effect of wind direction on the drag of a cylinder will also need to be considered as a realistic model of coniferous foliage. For cylinders with axes parallel to the flow Rouse (1978) gives the data for $Re > 10^3$ in Table A3.

Table A3 *Effect of l/d on Drag of Cylinders Parallel to Flow at High Reynolds Numbers*

l/d	0	1	2	4	7
C_D	1.12	0.91	0.85	0.87	0.99

As l/d increases, the drag coefficient passes through a minimum as form drag decreases and surface drag (or skin friction) increases with l for given d.

For cylinders at intermediate angles the situation is more complicated. As a first approximation we may consider resolving the wind vector into normal and parallel components (Figure A2).

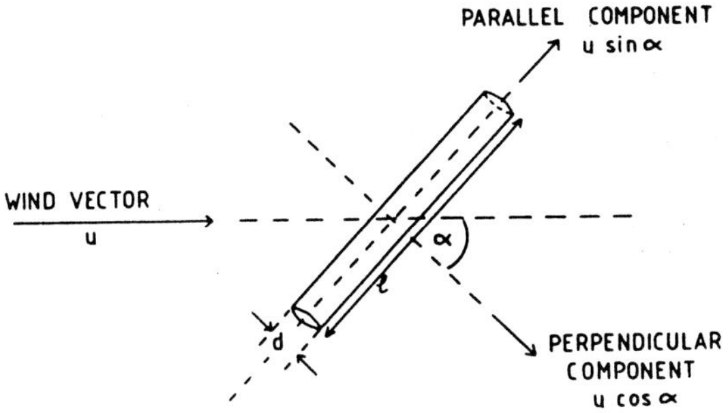

Fig A2. Normalisation of Flow into Parallel and Perpendicular Components.

The drag force for the two velocity components may then be calculated separately.

Normal Component

$$F_n = 0.5\,\rho u^2 \cos^2\alpha l d C_D \tag{A3}$$

when C_D is given by (A2) with $Re = ud\cos\alpha/\nu$.

Parallel Component

The drag coefficients in Table 3 are dominated by the ends of the cylinder. If instead one considers the curved surface, it is possible to consider the cylinder in terms of an equivalent flat sheet of length l and width πd provided the boundary layer thickness δ is much less than d. Schlichting (1964) gives an approximate expression

$$\delta = 5\left(\frac{\nu l}{u}\right)^{1/2} \tag{A4}$$

for the thickness of the laminar boundary layer at a distance l from the leading edge. Taking l = 17 mm and u = 1 m s^{-1} for needles gives a value $\delta \simeq 2.5$ mm which is larger than needle dimensions. Hence the approximation is likely to be a poor one. Nevertheless it will serve as an estimate of the order of magnitude. For the flat plate the Blasius solution gives

$$F_p = 0.664\,\pi\, d u^{3/2} \rho^{3/2} \nu^{1/2} l^{1/2} \sin^{3/2}\alpha \tag{A5}$$

The ratio of the two forces is

$$\frac{F_p}{F_n} = \frac{0.664\,\pi d u^{3/2} \rho^{3/2} \nu^{1/2} l^{1/2} \sin^{3/2}\alpha}{0.5\,\rho u^2\, l d C_D(ud\cos\alpha/\nu)\cos^2\alpha} \tag{A6}$$

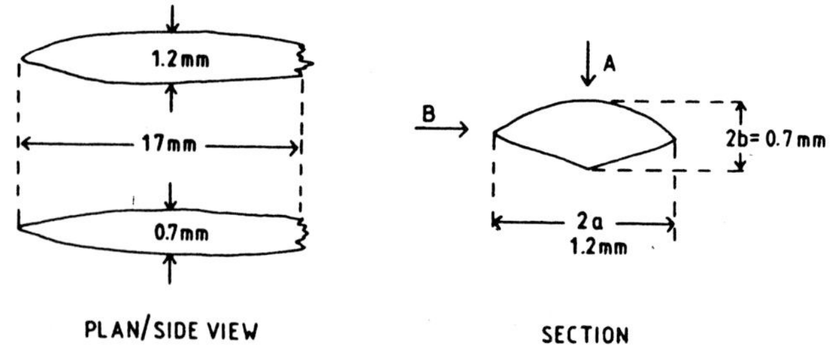

Fig A3. Cross-section of Needles.

If we assume that $4 < Re < 100$ in (A2)

$$C_D(u\cos\alpha d/\nu) = 5.5\,((ud\cos\alpha/\nu)^{-1/4}) \qquad \text{(A7)}$$

and take $\alpha = 45°$, $u = 1\ \mathrm{m\,s^{-1}}$ for needles

$$\frac{F_n}{F_p} \simeq 0.0003$$

Looked at another way, $F_p/F_n \simeq 1$ at an angle $\alpha = 89.44°$, hence the parallel component can be neglected.

The component of the drag force F_n in the direction of u is

$$F_n\cos\alpha = \tfrac{1}{2}\,\rho u\cos^3\alpha l d C_D(ud\cos\alpha/\nu) \qquad \text{(A8)}$$

So far the needles have been assumed to be cylindrical. In fact the cross-section and plan views are shown in Figure A3.

Some data on the drag of non-circular cylinders (particularly for elliptical cylinders and streamlined struts) are given by Goldstein (1965). For the wind direction "B" in Figure A3 the drag coefficient will be lower than for direction "A": the departure of the drag coefficient from that of a cylinder being greater the further the ratio of dimensions differ from 1.

As a rough guide for $Re \simeq 1000$, as a/b becomes large C_D will tend to 1.9 (flat plate at normal incidence) for direction "A", approximately twice as large as the drag on a cylinder at the same Reynolds number. The drag coefficient for direction "B" is $Re = 10^5$, $C_D \simeq 0.1$ a reduction by a factor of 10 relative to the value at $Re = 10^3$.

Treating needles as cylinders will therefore give a slight underestimate of drag when allowance is made for the different wind directions encountered in spruce shoots.

Analysis of Heat Transfer

The transfer of heat to and from an object of temperature T in a flowing fluid may be considered in terms of an effective boundary layer thickness for heat transfer

$$Q = \frac{k(T_S - T_A)}{\delta_Q} \tag{A9}$$

where Q is the average heat flux per unit area of the object. The ratio of a dimension of the object (d) to the thickness δ_Q is a dimensionless quantity usually referred to as the mean Nusselt number:

$$Nu = \frac{d}{\delta_Q} \tag{A10}$$

allowing equation A9 to be written

$$Q = k(T_S - T_A)Nu/d \tag{A11}$$

The Nusselt number will be a function of other dimensionless groups. In the case of forced convection, which dominates natural convection in our case, Nusselt number depends on Reynolds number.

Using cylinders as the simplest model for twigs and needles we may consider the data of Hilpert (1933) and others presented by McAdams (1954) reproduced in Figure A4 for cylinders at normal incidence.

The data in figure A4 can be represented by the expression

$$Nu = 0.32 + 0.43\,Re^{0.52} \quad \text{for} \quad (10^{-2} < Re < 10^3) \tag{A12}$$

Kramers (1946) gives a similar expression which when evaluated for air at 20° C gives

$$Nu = 0.39 + 0.51Re^{0.5} \quad \text{for} \quad (10^{-2} < Re < 10^4) \tag{A13}$$

Again the effect of the wind vector must be considered as in the case of drag. Fortunately, several workers have investigated the effect of yaw angle on the response of hot-wire anemometers and their results are relevant to the present discussion.

Referring to Figure A2 we may express the results in the form

$$u_{\mathrm{eff}}^2 = u^2(\cos^2\alpha + M^2\sin^2\alpha) \tag{A14}$$

where u_{eff} is the effective velocity for heat transfer. Webster (1962) obtained a value $M = 0.2$ independent of the l/d ratio for his hot wires but Champagne *et al* (1967) found that M did depend on l/d, being 0.2 at $l/d = 200$, but approximately zero at $l/d = 600$. Friehe & Schwartz (1968) found that their results were fitted by the relationship

$$u_{\mathrm{eff}}^2 = u^2(1 - N(1 - \cos^{1/2}\alpha))^4 \tag{A15}$$

where N varies from 1.0 for l/d approaching infinity down to 0.74 for

$l/d = 100$. In this formulation the equivalent M in equation A14 depends on α as well as l/d.

The origin of the effect of α on heat transfer (and perhaps the different experimental results) is that the hot wire does not have a uniform temperature along its full length owing to conduction to the supports. For our purposes we may follow Webster and take $M = 0.2$ and

$$u_{\text{eff}}^2 = u^2(\cos^2\alpha + 0.04\sin^2\alpha) \qquad \text{(A16)}$$

The value of Reynolds number is then

$$Re = \frac{u_{\text{eff}}d}{\nu}$$

and following Kramers (1946)

$$Nu = 0.39 + 0.51\,Re^{0.5}(\cos^2\alpha + 0.04\sin^2\alpha)^{0.25} \qquad \text{(A17)}$$

The effect of the non-cylindrical cross-section of needles on the heat transfer is difficult to assess. Jakob (1949) recommends calculating Reynolds number for a diameter such that the area of the exposed surface is the same as that of the non-circular cylinder and then gives expressions for various cross-sections. However the Reynolds numbers covered are those greater than 2000. In general the non-circular cylinders have higher Nusselt numbers than circular ones and hence have a better heat transfer capability.

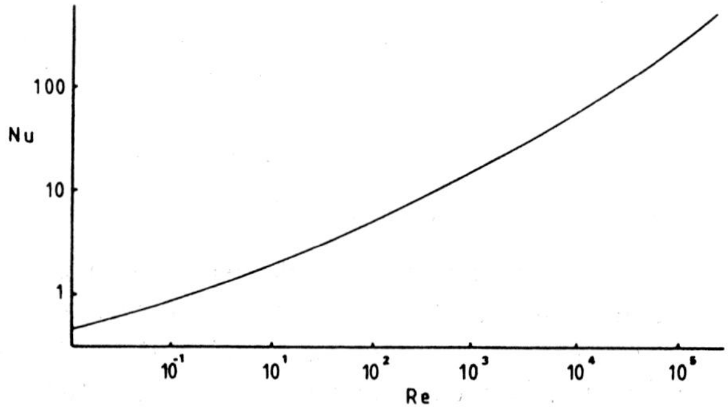

Fig A4. Data for Heating and Cooling Air Flowing Normal to a Single Cylinder, Corrected for Radiation to Surroundings. (McAdams 1954, p259.).

REFERENCES

Champagne, F.H., Sleicher, C.A. & Wehrmann, O.H. (1967). Turbulence measurements with inclined hot wires, Part I, Heat Transfer experiments with inclined hot wires. Journal of Fluid Mechanics, **28**: 153-175.

Freihe, C.A. & Schwartz, W.H. (1968). Deviations from the cosine law for yawed cylindrical anemometer sensors. Trans. ASME, J. Appl. Mech. **35E**: 655-662.

Goldstein, S. (1965). Modern Developments in Fluid Mechanics, Vol.II. New York: Dover Publications.

Jakob, M. (1949). Heat Transfer, Vol.I. New York: John Wiley & Sons.

Landsberg, J.J. & Powell, D.B.B. (1973). Surface exchange characteristics of leaves subject to mutual interference. Agricultural Meteorology, **12**: 169-184.

Landsberg, J.J. & Thom, A.S. (1971). Aerodynamic properties of a plant of complex structure. Quarterly Journal of the Royal Meteorological Society, **97**: 564-570.

McAdams, W.H. (1954). Heat Transmission. New York: McGraw-Hill.

Merzkirch, W. (1974). Flow Visualisation. London: Academic Press.

Monteith, J.L. (1973). Principles of Environmental Physics. London: Academic Press.

Rouse, H. (1978). Elementary Mechanics of Fluids. New York: Dover Publications.

Schlichting, H. (1968). Boundary Layer Theory. New York: McGraw-Hill.

Shuttleworth, W.J. (1976). One dimensional theoretical description of the vegetation-atmosphere interaction. Boundary-Layer Meteor. **10**: 273-302.

Thom, A.S. (1968). The exchange of momentum, mass and heat between an artificial leaf and the airflow in a wind tunnel. Quarterly Journal of the Royal Meteorological Society, **94**: 44-53.

Webster, C.A.G. (1962). A note on the sensitivity of a hot wire anemometer. Journal of Fluid Mechanics, **13**: 307-312.

List of Symbols

A	Characteristic area of body projected in a plane perpendicular to local or mean flow vector.
A_i	Projected area of individual shoots.
A_N	Projected area of needles.
A_T	Projected area of twig.
A_S	Projected area of shoot.
C_D	Coefficient of drag defined $F = \frac{1}{2}\rho u^2 A C_D$
C_{Di}	Coefficient of drag for individual elements.

C	Transfer coefficient, experimentally found.
C'	Calculated coefficient of transfer.
D	Twig diameter.
D_e	Effective diameter of shoots.
F	Force acting on shoot in direction of flow vector.
F_n	Drag force on needle caused by the normal component of wind.
F_p	Drag force on needle caused by parallel component of wind.
F_a	Force calculated for unsheltered shoot.
F_b	Force calculated for a sheltered shoot.
L	Length of twig.
M	Coefficient used by Webster (1962).
N	Coefficient used by Friehe & Schwartz (1968).
Nu	Nusselt number, $Nu = d/\delta$.
Q	Heat transfer rate per unit area.
Q_a	Heat transfer rate per unit area calculated for unsheltered shoot.
Q_b	Heat transfer calculated for sheltered shoot.
Re	Reynolds Number, $Re = du/\nu$.
S	Surface area.
T_S	Temperature of surface.
T_A	Ambient air temperature.
a	Length of the major axis of elliptical cross-section needles.
b	Length of the minor axis of elliptical cross-section needles.
d	Representative diameter of needle.
l	Length of needle or cylinder.
k	Thermal conductivity of air.
n	Number of needles.
p	Shelter factor according to Landsberg & Thom (1971).
p_c	Calculated shelter factor.
r	Distance from twig in the r-direction of cylindrical polar coordinates.
u	Wind velocity.
u_o	Wind velocity at an infinite distance upstream of the shoot.
u_n	Normal component of wind velocity.
u_p	Parallel component of wind velocity.
$(u_n)_o$	Component of u resolved in the direction of undisturbed flow.

u_{eff}	Effective wind velocity.
u_r	Radial wind velocity.
α	Angle between local wind vector and axis of needle.
δ	Thickness of boundary layer for momentum transfer.
δ_Q	Thickness of boundary layer for heat transfer.
θ	Angle subtended by needle and axis of twig.
ν	Kinematic viscosity, for air at 20° C = 0.15×10^{-4} $m^2 s^{-1}$.
ρ	Density of fluid, 1.2 kg m^{-3} for air at 20° C.
σ	Shoot density factor $(A_N + A_T)/A_S$.
ϕ	Angle subtended between needle and flow vector.

Progress in Biometeorology, 1985, Vol. 2, pp. 37-48

A METHOD OF DETERMINING THE THERMAL RESISTANCE OF POIKILOTHERMS FROM A MODEL OF HEAT EXCHANGE IN AIR AND WATER

C.J. Bell
Glasshouse Crops Research Institute,
Worthing Road,
Littlehampton,
West Sussex, BN17 6LP.

INTRODUCTION

All animals exchange heat with their environment depending on whether they are warmer or cooler than their surroundings. The rate of exchange is a function both of the physics of heat transfer and of the mechanisms used by the animal to increase or decrease this transfer of energy. The latter may include behavioural responses, the control of circulation and erectile fur or feathers for example. As heat is exchanged, the animal's body temperature will tend to change, but homeotherms will minimise this by adjusting the rate of endogenous heat production. Poikilotherms have no mechanism for thermogenesis so that their body temperatures will follow that of the environment.

When radiation can be neglected, the physics of heat exchange can be encompassed in the concept of *thermal resistance* (Monteith, 1973). The heat flux Φ is given by the Ohm's Law analogue

$$\Phi = \rho c_p(T_a - T_c)/r_H \qquad 1)$$

where ρ and c_p are the density and specific heat of air, T_a and T_c are the ambient and body-core temperatures, and r_H is the thermal resistance. When resistance is defined in this way it has units of s m^{-1} and can be compared with resistances to momentum and mass transfer (Clark *et al*, 1973).

To estimate r_H we must measure Φ, T_a and T_c. In homeotherms T_c is relatively constant so, by maintaining T_a constant, only Φ need be measured. In poikilotherms, T_c is constant only when $\Phi = 0$, and we must measure all three variables to estimate r_H. This task is difficult experimentally and an alternative approach has been used for lizards.

Experimentally, the lizard is restrained in a uniform environment and heating and cooling curves produced by imposing changes in the environmental temperature and following the change in body temperature.

If the animal can vary its thermal conductance, or produce heat endogenously, the heating or cooling curve will differ from that of a passive body. The rates of heating and cooling are quantified as *thermal time constants* (Spotila *et al*, 1973; Smith, 1976; Grigg *et al*, 1979). A model of heat exchange is then used to interpret the experimental thermal time constants. The models mimic the structural complexity of the reptile to a greater (Spotila *et al*, 1973) or lesser (Grigg *et al*, 1979) extent. The physics of heat exchange is treated similarly, with the result that the models contain a large number of free parameters. Even a simple approximation given by the latter authors involves 7 independent parameters.

It was therefore considered appropriate to construct a simple model with a minimum number of variables, to determine if such a model could account for the differences in thermal time constant observed for various lizards. An outline of this model has been given previously (Bell, 1980), and compared with data from the literature on the heating and cooling of lizards. Here, I wish to consider the mathematical details of the model; to give some solutions for non-steady-state environments, to consider the validity of the assumptions of the model and to show how the thermal resistance of lizard skin can be found from a comparison of heat exchange in air and water.

The derivation of the model is by no means new, as it expresses the basic relation of heat conduction that the flux of heat is proportional to the temperature difference. This law (called either Newton's Law of Cooling or Fourier's Law of Heat Flow), has often been applied to animals (see, for instance, Strunk, 1971; Tracy, 1971; Kleiber, 1972; Monteith, 1973; Porter *et al*, 1973; or Bakken, 1976). However, few authors have considered the non-steady state, exceptions being Porter *et al* (1973) who considered the cooling of a lizard in a wind tunnel; Buatois & Croze (1978), who looked at the thermal response of the cockroach; and Monteith & Butler (1979), who developed a model of thermal lag and subsequent dew formation on cocoa pods.

THE MODEL

Consider a body of arbitrary shape consisting of an isothermal core surrounded by a thin insulating layer of negligible heat capacity in an isothermal environment. If we define T_a as the ambient temperature, T_c as the core temperature, h $(=1/r_H)$ as the thermal conductance of the insulating layer and A the area of the body, the amount of heat transferred to the body per unit time is given by

$$\frac{dQ}{dt} = hA(T_a - T_c). \qquad 2)$$

If the heat capacity of the core is C, its rate of temperature increase will be $1/C \times dQ/dt$, and so,

$$\frac{dT_c}{dt} = \frac{hA}{C}(T_a - T_c). \qquad 3)$$

This is a linear, first-order differential equation in T_c.

In general, T_a is a particular function of space and time, varying as the animal moves about and the ambient temperature fluctuates. In certain simple cases, the solution of the above equation can be found, and three of these are presented below. However, it is useful to investigate the general behaviour of the solutions first. Consider the homogeneous equation,

$$\frac{dT_c}{dt} = -\frac{hA}{C}T_c. \qquad 4)$$

This has the solution

$$T_c(t) = ke^{(-hA/C)t}, \qquad 5)$$

where k is a constant. Thus, by the Principle of the Complementary Function, every solution of equation (3) can be written

$$T_c(t) = f(T_a,t) + ke^{(-hA/C)t}, \qquad 6)$$

where $f(T_a,t)$ is a particular solution of equation (3). Therefore every nontrivial solution contains an exponential term with exponent $(-hA/C)t$ and, in accordance with common practice, we set this equal to $-t/\tau$, where $\tau = C/hA$ and is called the *thermal time constant*.

PARTICULAR SOLUTIONS

Particular solutions of equation (3) can be found for the following simple boundary conditions.

Step function

Consider a step-function change in ambient temperature;

$$T_c = T_a = T_0 \qquad t < 0$$

and

$$T_a = T_1 \qquad t > 0.$$

In this case, a change of variable to $\Delta T = T_c - T_a$ yields the solution

$$\Delta T(t) = \Delta T(0)e^{-t/\tau},$$

where $\Delta T(0) = T_1 - T_0$. This is the situation most commonly encountered in heating and cooling curve experiments, and τ can be found from the slope of the graph of $\log(\Delta T)$ plotted against time.

Ramp function

Another interesting case is that of a ramp-function change in ambient temperature,

$$T_a = T_0 + \alpha t,$$

where the ambient temperature increases or decreases linearly with time. Equation (3) can now be written

$$\frac{dT_c}{dt} = \frac{1}{\tau}(T_0 + \alpha t - T_c). \qquad 7)$$

It is evident that $T_c = T_0 + \alpha t - \alpha\tau$ satisfies this equation, giving a general solution of

$$T_c = T_0 + \alpha t - \alpha\tau + ke^{-t/\tau}.$$

Applying the initial condition that $T_c(0) = T_0$ gives $k = \alpha\tau$, so that the particular solution is

$$T_c = T_0 + \alpha t - \alpha\tau(1 - e^{-t/\tau}), \qquad 8)$$

or

$$\Delta T = -\alpha\tau(1 - e^{-t/\tau}). \qquad 9)$$

After a period much greater than τ, this becomes a constant temperature difference of $-\alpha\tau$. This solution has been applied to temperature changes in the body of the cockroach by Buatois & Croze (1978). They found good agreement between the derived time constants and those from a step-function change in ambient temperature. Temperature variations in the field often approximate ramp functions, and this solution has been applied by Boland & Bell (1980) to calculate the thermal time constants of crocodiles in an outdoor enclosure.

Sinusoidal variations

The diurnal variation of ambient temperature is often approximately sinusoidal, so it is of interest to consider the function,

$$T_a = T_0 + T_1 \sin \omega t$$

where T_0 is the mean temperature of the environment and T_1 is the amplitude and ω the angular frequency of the sinusoidal variation in temperature. If we assume that the core temperature changes likewise, but with a different amplitude and phase,

$$T_c = T_0 + T_b \sin(\omega t - \phi)$$

we find that,

$$\tan \phi = \omega \tau$$

and

$$T_b = T_1 \cos \phi.$$

Thus, when the time constant is equal to half the period (e.g. a time constant of 12 hours for a 24 hour day), the core temperature will lag the environmental temperature by $\tan^{-1}(2\pi/2)$ or 72°, and will vary by only 30% of the ambient fluctuation. Similar phase lags and a decrease in amplitude are seen in the numerical simulations of Spotila *et al* (1973) when considering diurnal changes in the environment.

THE ASSUMPTIONS

It is most important to note the assumptions of this model, and to ask if they are reasonable. These are:

(i) The core is isothermal,
(ii) The insulating layer is thin and of negligible heat capacity,
(iii) Thermal conductance h and heat capacity C are constant,
(iv) Endogenous heat production is negligible,
(v) The environment is isothermal.

Assumptions (i) – (iv) all relate to the 'animal' being modelled and essentially specify the applicability of the model. Any real animal will be an approximation to this ideal. Assumption (i) has been tested for crocodiles by Drane *et al* (1977). They measured temperatures in various parts of the body of a crocodile as it heated in sunshine. As the cloacal temperature rose from 20°C to 30.5°C, the difference between sites in the core was at worst 3.3 K, and averaged only 1.8 K. Thus assumption (i) is not unreasonable, at least as a first assumption.

Assumption (ii) is also reasonable, as the skin and subcutaneous fat will constitute the insulating layer. The constance of h (assumption iii) is not guaranteed, and indeed this is the parameter that is varied by the animal in attempting to maintain thermostability. However, under thermal stress, it can be assumed to be held at its maximum or minimum value (Boland & Bell, 1980). The thermal capacity C is a property of the material of the core, and would not be expected to vary unless the animal was severely dehydrated.

Endogenous heat production (assumption iv) in poikilotherms is usually assumed to be small, and so would cause only a small rise in body temperature. Results given by Bakken (1976) suggest that this was less than 1 K for the lizard *Sceloporus occidentalis* cooling in air at about 25°C. However, it has been suggested that even a small temperature increase could seriously affect an analysis using the ramp function result, equation (9) (Drane, pers. comm.).

Assumption (v) relates to the environment, and gives the condition under which the model can be used. It will be satisfied for an animal heating or cooling in well-mixed water where the thermal resistance of the boundary layer is small. In air, the thermal resistance of the boundary layer will be comparable to that of the body, and can be considered as part of the insulating layer. It is then possible to consider that an animal in a stream of air will satisfy this assumption, if radiation is negligible.

The last condition is important, and should be considered in more detail. Radiative heat loss from a body follows the Stefan-Boltzmann equation:

$$\frac{dQ}{dt} = \varepsilon\sigma A T_b^4 \qquad 10)$$

where ε is the emissivity of the surface, A the area, T_b the surface temperature of the body and σ the Stefan-Boltzmann constant (5.67×10^{-8} W m^{-2} K^{-4}). The body also absorbs heat from its surroundings, which are radiating at their own temperature, T_a. Thus, the net loss (or gain) of heat will be

$$\frac{dQ}{dt} = \sigma A(\varepsilon T_b^4 - \alpha\varepsilon_1 T_a^4) \qquad 11)$$

where α is the absorbtivity of the body and ε_1 is the emissivity of the surroundings. This equation can be linearized over a limited range of temperatures by factoring as follows:

$$\frac{dQ}{dt} = \sigma A\varepsilon[T_b - \left(\frac{\alpha\varepsilon_1}{\varepsilon}\right)^{1/4} T_a].[T_b + \left(\frac{\alpha\varepsilon_1}{\varepsilon}\right)^{1/4} T_a].[T_b^2 + \left(\frac{\alpha\varepsilon_1}{\varepsilon}\right)^{1/2} T_a^2].$$

This can then be added as a parallel path of flow to the losses by conductance and convection. However, for small temperature differences, the net flow of heat by thermal radiation is often much smaller than that by conduction and convection, and can be ignored. This will always be true for an animal in water, but will also often apply in fast-moving air, as will now be shown.

In many of the cooling-curve experiments on lizards, a step function change of 10 K is imposed, say from 30°C to 20°C. Assuming $\varepsilon = \varepsilon_1 = \alpha = 0.95$, we can see that, at the start of the experiment, the rate of heat loss by radiation (dQ/dt) is 77 W m^{-2}. At an air velocity of 3 m s^{-1}, the convective heat loss from a typical reptile under these conditions (see later) would be of the order of 500 W m^{-2}. Thus radiative heat loss is small compared with convective heat loss, and can be ignored.

In the field, however, radiation temperatures will often differ greatly from body temperature, as when an animal is exposed to sunlight or to a clear sky, and radiation will then be of overriding importance.

SCALING BEHAVIOUR

The prediction of the model that the thermal time constant is equal to

C/hA is not open to direct test, because these variables cannot be measured. However, in the literature there are many measurements of thermal time constant on several species of lizard over a wide range of size. It is therefore of interest to look at the scaling behaviour of τ.

It is easily shown (e.g. Gold, 1977, p 54) that the area of a body of constant shape varies as the linear dimension (d) squared, and the volume as d^3. Thus C, which is proportional to the mass M, will also scale as d^3. The thermal conductance h is inversely proportional to the thickness of the insulating layer which is assumed to scale as d. Thus the thermal conductance will scale as d^{-1}, and $\tau = C/hA$ will scale as d^2, or as $M^{2/3}$.

This proposition has been tested in an earlier paper (Bell, 1980), and shown to agree with the data for cooling in water, but not with that for heating in water (Figure 1). This result contradicts that of Kleiber (1972), who found that thermal conductance scaled as $M^{-1/2}$. This would imply that

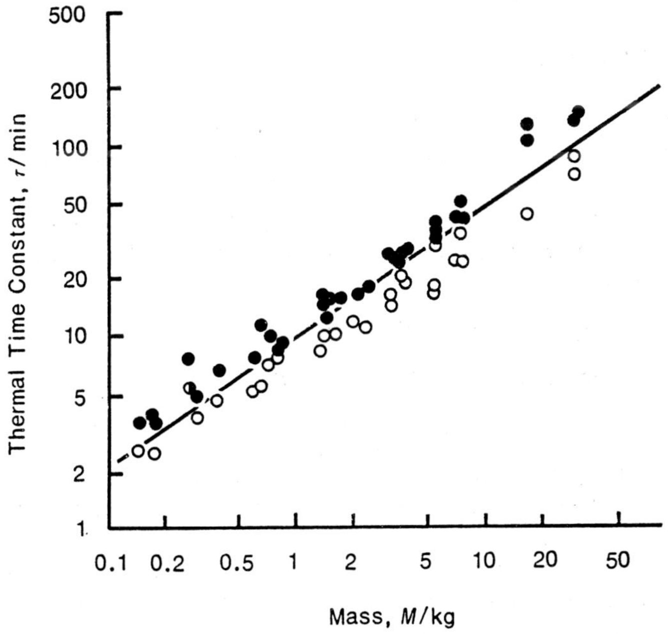

Fig 1. Log – log plots of thermal time constant against mass for various lizards heating (open symbols) and cooling (filled symbols) in water. The data have been taken from the literature (for details see Bell, 1980). A line of slope 0.67 as predicted by the model is shown for comparison.

thermal time constant scales as $M^{1/2}$, which is clearly inconsistent with the data of Figure 1. Kleiber derives his results solely from data on birds and mammals, and lizards may not follow the same allometry, perhaps because they are poikilotherms whereas birds and mammals are homeotherms and maintain body temperature by endogenous heat production.

Note that there is now an additional assumption in the model. We have assumed that the thickness of the insulating layer is proportional to the linear dimension of the animal. That is, that the "shape" of the animal is the same over the whole size range or that the series of animals is allotropic. This also implies that the physiological mechanisms inherent in thermoregulation are active over the whole size range (see also Smith, 1979). Perhaps the disagreement for heating could be due to the breakdown of this latter assumption in that the physiological mechanisms in larger animals could be geared more to enhancing heat uptake than reducing heat loss, thus reducing the thermal time constants below those expected for large lizards heating in water (Figure 1).

HEAT EXCHANGE IN AIR

Engineering studies of heat exchange in air conventionally use non-dimensional groups to analyse the effects of size and windspeed. Thus, the Reynolds number (*Re*) characterizes the velocity distribution of the boundary layer, and the Nusselt number (*Nu*) gives an analogous description of the temperature (see, for example, Ede, 1967 or Monteith, 1973). For forced convection, these are related by means of the Prandtl number (*Pr*), so that

$$(Nu) = (Re)^n (Pr)^m$$

where n and m are constants. In air, *Pr* is a constant, so that

$$(Nu) = C_h (Re)^n. \qquad 12)$$

C_h and n are constants for bodies of the same shape, and have been determined for many regular shapes such as spheres, cylinders, flat plates etc.

The calculation of both *Re* and *Nu* depends on the choice of some characteristic dimension of the body, d. For regular bodies this causes little difficulty, and d is chosen as the diameter of a sphere or cylinder or the distance along a flat plate. However, for irregular shapes, such as animal bodies, the characteristic dimension must be chosen more carefully. Mitchell (1976) has shown that an appropriate dimension for many animals is the cube root of the volume. When *Nu* is plotted against *Re*, both calculated using this characteristic dimension, the lines from many animals ranging from cows to frogs and flying insects fall closely together and also close to that for a sphere. However, the results quoted for lizards do not show the same regularity, which suggests that there may be a more appropriate characteristic dimension for these animals.

Although the Reynolds and Nusselt numbers are very important in engineering studies of heat flow, they may not be appropriate in studying the heat flow from poikilothermic animals. The derivation of both numbers involves the dimension of the body and the windspeed, thus confounding these variables. If the scaling of the body does not follow a simple pattern, it will lead to inconsistencies. As the relevant variable is the heat loss from the animal, it is better to use the direct measure of this, the resistance to heat flow r_H.

We can now treat the insulating layer around the core of our animal as consisting of two components, one being the layer of skin and fat, and the other the boundary layer of air next to the skin, as has been done by Tracy (1971), Kleiber (1972), Porter *et al* (1973), Bakken (1976) and Bell (1980). In the last paper it was shown that the component of the thermal time constant due to the boundary layer could be calculated by subtracting the thermal time constant found in water from that in air. The resulting time constants (Figure 2) are similar whether the lizard was heating or cooling, unlike the complete time constants which show marked differences between heating and cooling in air (Grigg *et al*, 1979). This is consistent with the

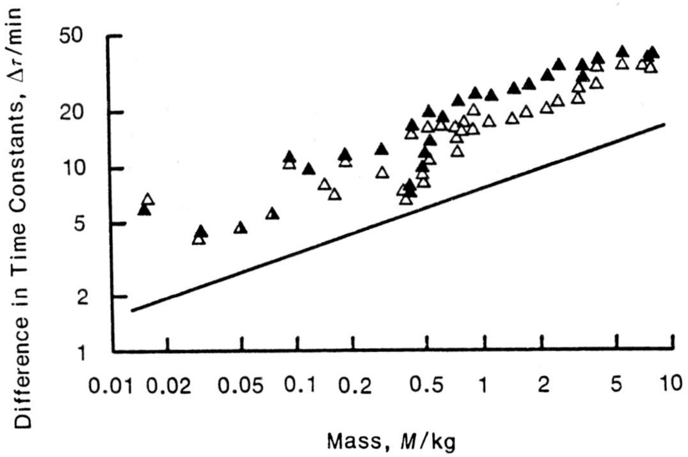

Fig 2. Log – log plots of the component of thermal time constant in air due to the boundary layer, against mass for both heating (open symbols) and cooling (filled symbols). The calculation of this component has been described previously (Bell, 1980). A line of slope 0.33, as shown, is predicted if the boundary layer resistance is constant.

assertion that this component represents solely the thermal resistance of the boundary layer and hence should be the same for both heating and cooling.

The boundary layer component of the thermal time constant appears to scale as $M^{1/3}$ (Figure 2), implying that the resistance due to the boundary layer is approximately constant for lizards with masses ranging from 30 g to 8 kg, at a constant windspeed of 3 m s^{-1}. Bell (1980) suggested that this was a consequence of the roughness of lizard skin, and that the appropriate linear dimension of a lizard may be related more to the characteristics of the skin than to the overall size of the animal. This is consistent with the results of Mitchell (1976) mentioned earlier, who found diffeences between lizards and smooth-skinned species such as cows, sheep, frogs and men.

We now have a relation between thermal time constant and boundary layer resistance, although the latter is specified only as that of a lizard of given mass. We wish to calibrate the boundary layer resistance by estimating it from the Nusselt number (Monteith, 1973), and hence from the air velocity and characteristic dimension of the body. If we take 1 cm to be the characteristic dimension of surface roughness of lizard skin, the Reynolds number is 2000 at the windspeed of 3 m s^{-1} used in Bell (1980). Taking values of C_h and n in equation (12) for forced convection from cylinders, the corresponding Nusselt number is 22. The boundary layer resistance of a lizard can now be calculated to be approximately 21 s m^{-1}.

We now consider the relationship between thermal time constant and boundary layer resistance for a lizard with boundary layer resistance r_H. This has been shown to be

$$\tau = kM^{1/3} r_H$$

where k is a proportionality constant and M the mass of the body (Bell, 1980). Choosing an arbitrary mass of 1 kg, and reading from the graph of thermal time constant due to the boundary layer against mass (Figure 2), then this time constant is approximately 17 min for an air velocity of 3 m s^{-1}. Thus

$$\tau \simeq 0.8\, r_H$$

at a mass of 1 kg when τ is in minutes and r_H is in s m^{-1}.

Having determined an approximate relationship between thermal time constant and resistance from the respective experimental and theoretical values of these for the boundary layer, we wish to determine the resistance of lizard skin. To do this we take the experimental values of thermal time constant for heating and cooling in water for a 1 kg lizard. These are 8 and 11.5 minutes respectively (Figure 1). Using the relation above, we get resistances of 10 and 14 s m^{-1}. Because direct measurements of tissue thermal resistance have usually been made on domestic animals and man, we scale the resistances up to a 50 kg animal, giving 37 and 52 s m^{-1} for heating and cooling respectively. These can be compared with the values of about 50 and 120 s m^{-1} given by Monteith (1973, p 122) for vaso-dilated and -constricted tissue of various mammals.

REFERENCES

Bakken G.S. (1976). An improved method for determining thermal conductance and equilibrium body temperature with cooling curve experiments. Journal of Thermal Biology **1**: 169-175.

Bell, C.J. (1980). The scaling of the thermal inertia of lizards. Journal of Experimental Biology **86**: 79-85.

Boland, J.E. & Bell, C.J. (1980). A radiotelemetric study of heating and cooling in unrestrained, captive *Crocodylus porosus*. Physiological Zoology **53**: 270-283.

Buatois, A. & Croze, J.P. (1978). Thermal responses of an insect subjected to temperature variations. Journal of Thermal Biology **3**: 51-56.

Clark, J.A., Cena, K. & Monteith, J.L. (1973). Measurements of the local heat balance of animal coats and human clothing. Journal of Applied Physiology **35**: 751-754.

Drane, C.R., Webb, G.J.W. & Heuer, P. (1977). Patterns of heating in the body, trunk and tail of *Crocodylus porosus*. Journal of Thermal Biology **2**: 127-130.

Ede, A.J. (1967). An Introduction to Heat Transfer Principles and Calculations. Oxford: Pergamon.

Gold, H.J. (1977). Mathematical Modelling of Biological Systems. New York: Wiley.

Grigg, G.C., Drane, C.R. & Courtice, G.P. (1979). Time constants of heating and cooling in the Eastern Water Dragon, *Physignathus lesueurii* and some generalizations about heating and cooling in reptiles. Journal of Thermal Biology **4**: 95-113.

Kleiber, M. (1972). Body size, conductance for animal heat flow and Newton's law of cooling. Journal of Theoretical Biology **37**: 139-150.

Mitchell, J.W. (1976). Heat transfer from spheres and other animal forms. Biophysics Journal **16**: 561-569.

Monteith, J.L. (1973). Principles of Environmental Physics. London: Arnold.

Monteith, J.L. & Butler, D.R. (1979). Dew and thermal lag: a model for cocoa pods. Quarterly Journal of the Royal Meteorological Society **105**: 207-215.

Porter, W.P., Mitchell, J.W., Beckman, W.A. & DeWitt, C.B. (1973). Behavioural implications of mechanistic ecology. Oecologia **13**: 1-54.

Smith, E.N. (1976). Heating and cooling rates of the American Alligator, *Alligator mississippiensis*. Physiological Zoology **49**: 37-48.
Smith E.N. (1979). Behavioural and physiological thermoregulation of crocodilians. American Zoology **19**: 239-247.
Spotila, J.R., Lommen, P.W., Bakken, G.S. & Gates, D.M. (1973). A mathematical model for body temperatures of large reptiles: implications for dinosaur ecology. American Naturalist **107**: 391-404.
Strunk, T.H. (1971). Heat loss from a Newtonian animal. Journal of Theoretical Biology **33**: 35-61.
Tracy, C.R. (1971). Newton's law: its application for expressing heat losses from homeotherms. Bioscience **22**: 656-659.

Progress in Biometeorology, 1985, Vol. 2, pp. 49-58

HEAT LOSS AND THE THERMAL ENVIRONMENT OUTDOORS

A.J. McArthur
Department of Physiology and Environmental Science,
University of Nottingham School of Agriculture,
Sutton Bonington,
Loughborough,
LE12 5RD, U.K.

INTRODUCTION

Thermal strain imposed by the climatic environment reduces the productivity of livestock. In cold, if metabolic rate is raised to prevent a drop in body-core temperature, then a higher proportion of the food intake is employed in thermoregulation and body mass declines if the metabolisable energy intake does not meet the energy demands of the environment. In heat, productivity is lowered when elevated body temperatures and high respiratory rates reduce an animal's appetite and, hence, its food intake. If we could estimate the rate of heat loss from livestock and their thermoregulatory responses in relation to the climatological variables, more exact recommendations could be framed for the control of these variables to minimise thermal strain and improve productivity. Furthermore, comparison of the response of different animals to the same climatic environment would help to guide the selection of breeds tolerant to stress.

The climatological variables which determine the heat balance of livestock include temperature, windspeed, humidity and radiation. These variables are easily measured and records for different climates are available. However, it has proved difficult to assess the thermal status of livestock exposed to different combinations of these variables, largely because of the complexity of the interaction. This paper describes the basic principles involved in this interaction, and outlines the development of an equation to evaluate the heat flow from an animal's body in a specified environment. This equation incorporates the physical and physiological parameters which influence the flow of heat and water vapour between the animal and its environment. The paper illustrates how this analysis can be applied to estimate the energy requirements of cattle in cold, and predict their body-core temperature and respiratory rate in heat.

BASIC CONSIDERATIONS

The maintenance of a relatively constant body temperature by a homeothermic animal depends on the partitioning of available energy (from metabolism and net radiation) between sensible and latent heat (Monteith, 1981). In a thermoneutral environment the rate at which heat is produced by metabolism ranges from about 50 to 200 W m^{-2}, depending on species and the level of production (Graham *et al*, 1959; Webster, 1974). In strong sunshine outdoors, the rate at which heat is gained by absorption of solar radiation can exceed the metabolic rate by a factor of 3 or 4 (Finch, 1972). Sensible heat is lost to the environment by thermal radiation exchange and convection (Kelly *et al*, 1954; Wiersma & Nelson, 1967). Heat is also dissipated to the surroundings by evaporation from the skin surface and respiratory system (Hales, 1974). For an animal, the flux densities of sensible and latent heat, as well as metabolism, can be based on body surface area.

The steady flux density of *sensible* heat Q (W m^{-2}) through an insulating layer with thermal resistance r (s m^{-1}) can be expressed as

$$Q = \rho c_p(\delta T)/r \qquad 1)$$

where δT (K) is the temperature difference across the layer, and ρc_p (= 1.29 x 10^3 J m^{-3} K^{-1}) is the volumetric specific heat of air at an arbitrary temperature of 0 °C (Monteith, 1975). A resistance r=1.0 s cm^{-1} (100 s m^{-1}) is equivalent to an insulation of 0.078 K m^2 W^{-1} (Cena & Clark, 1978).

The corresponding relation for *latent* heat transfer λE (W m^{-2}) from a wet surface is given by

$$\lambda E = \rho c_p(\delta e)/\gamma r_v \qquad 2)$$

where λ (J g^{-1}) is the latent heat of vaporization of water, and E (g m^{-2} s^{-1}) is the rate of water loss per unit area by evaporation. The quantity δe (mbar) is the difference in vapour pressure between the surface and the surrounding air, γ (= 0.66 mbar K^{-1}) is the psychrometer constant, and r_v (s m^{-1}) is the resistance to water vapour transfer.

THEORY

The heat transfer between an animal and its environment can be described by a simple resistance analogue (e.g. Bakken & Gates, 1975). The resistance analogue in Figure 1 describes the flow of sensible and latent heat from an animal with a body-core temperature T_b to an environment specified by the air temperature T_a, mean radiative temperature $\overline{T}_e$, and vapour pressure e_a. The quantity $\overline{S}$ (W m^{-2}) is the mean solar irradiance over the animal's body, and ρ_c is the mean reflection coefficient of the coat. The mean thermal resistance of the body tissue is indicated by $\overline{r}_s$, the mean boundary-layer resistance by $\bar{r}_a$, and r_R is the resistance to thermal radiation exchange with the environment. The resistance $\bar{r}_{vs}$ to water vapour transfer from the skin surface to the surrounding air is provided by the coat and boundary-layer (acting in series), and r_{vr} is the resistance to water vapour loss from the respiratory system. The resistance of the coat to sensible heat transfer is denoted by $\bar{r}_c$. The quantities C and L_n are the flux densities of sensible heat from the coat surface by convection and thermal radiation exchange respectively, and λE_s and λE_r are the flux densities of latent heat from the skin surface and respiratory system respectively. Sensible heat loss from the respiratory system is usually negligible. The rate of heat flow through the body tissue is represented by G_s.

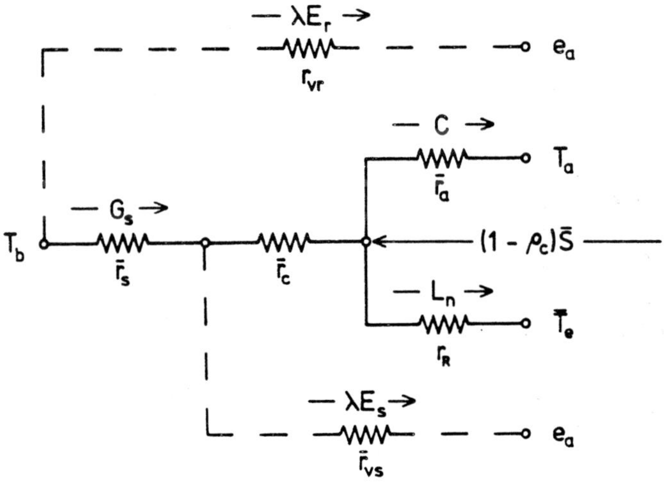

Fig 1. Resistance analogue for animal heat loss (symbols defined in the text).

The heat flow G_s is of basic significance in an animal's total heat balance. When G_s *exceeds* the rate at which heat is produced by metabolism, then metabolic rate must rise to prevent a drop in body temperature. Conversely, when G_s is *less* than the rate of heat production, then the rate of respiratory heat loss λE_r must increase to prevent a rise in body temperature.

Evaluation of the heat flow G_s will, therefore, provide insight into the energy balance and thermal status of an animal in a given environment.

The transfer of heat, or water vapour, across each resistance component shown in Figure 1 (apart from the coat resistance $\bar{r}_c$) can be described by equation (1) or (2). The different equations can then be combined to derive an expression for the heat flow G_s, and from which terms that include the temperature or vapour pressure at the skin and coat surface have been eliminated. The analysis is complicated by the penetration of an animal's coat by solar radiation: some of the incident radiation may be absorbed at the skin surface as well as throughout the whole depth of the coat. Provided the physical characteristics of the coat are known, the distribution of radiation absorption between the skin and coat surface can be calculated. For the analogue presented in Figure 1, the heat flow G_s can be evaluated from the equation

$$G_s = \frac{\rho c_p(\theta'_b - \theta'_a) - \bar{r}_{aR} R_{ni} + [\bar{r}_{co} t/\bar{l}][(1-\rho_c)\bar{S} - 1/t \int_0^{\bar{l}-t} S(z)(\bar{l}-z)dz - \int_{\bar{l}-t}^{\bar{l}} S(z)dz]}{\bar{r}_s[1 + ((\bar{l}-t)\bar{r}_{co}/\bar{l}\bar{r}_s) + \bar{r}_{aR}/\bar{r}_s + \Delta_b/\gamma']} \qquad 3)$$

As expected, equation (3) reveals that the rate of heat loss from the body surface depends on physical and physiological factors: the physical factors include the climatological variables and the resistances to heat and mass transfer between the skin surface and the environment; the physiological factors include the regulation of tissue resistance, sweat rate and body temperature. The magnitude of the resistance components in equation (3) and their dependence on physical factors was reviewed by McArthur (1981). The physiological factors relevant to the present analysis, and details of the derivation of equation (3), will be published in a later paper.

Equation (3) has a unique solution when an animal whose thermoregulatory responses are known is exposed to a given combination of climatological variables. Provided the thermoregulatory characteristics are related to the physical state of the animal itself, the heat flow G_s can be determined without prior knowledge of this state, i.e. only the environmental variables and the animal characteristics need be specified. The derivation of major variables in equation (3) will now be reviewed.

Apparent Equivalent Temperature θ'

The apparent equivalent temperature of an animal θ'_b is defined by

$$\theta'_b = T_b + e_s(T_b)/\gamma' \qquad 4a)$$

where $e_s(T_b)$ is the saturation vapour pressure (s.v.p.) at body-core temperature T_b. The modified psychrometer constant γ' accounts for incomplete wetness of the skin surface and the resistance to heat and water vapour transfer provided by the coat and boundary-layer. The value of γ' can be determined from

$$\gamma' = \gamma \bar{r}_{vs}/\omega(\bar{r}_{aR}+\bar{r}_c) \qquad 5)$$

where $\bar{r}_{aR} = \bar{r}_a r_R/(\bar{r}_a+r_R)$ is the combined resistance to sensible heat loss by convection and thermal radiation exchange between the outer surface of the body and the environment. The quantity ω is given by

$$\omega = \lambda E_{sd}\gamma \bar{r}_{vs}/\rho c_p[e_s(\bar{T}_s)-e_a] \qquad 6)$$

where E_{sd} is the rate of sweat deposition at the skin surface, $e_s(\bar{T}_s)$ is the s.v.p. at mean skin temperature $\bar{T}_s$, and $\bar{r}_{vs}$ is the resistance to water vapour transfer provided by the coat and boundary-layer when the skin surface is completely wet, i.e. ω is the ratio of the actual rate of evaporation of sweat to the rate which would occur (potential evaporation rate) if the skin surface was completely wet. When the rate of sweat deposition E_{sd} exceeds the potential evaporation rate, then ω = 1.0. However, when E_{sd} is less than the potential evaporation rate, then $\omega < 1.0$.

The apparent equivalent temperature of the air θ'_a is defined by

$$\theta'_a = T_a + e_a/\gamma' \qquad 4b)$$

Thus, for example, when $\bar{r}_{vs}$ = 3.0 s cm^{-1}, $\bar{r}_a$ = 3.0 s cm^{-1}, r_R = 2.0 s cm^{-1}, $\bar{r}_c$ = 0.7 s cm^{-1} and ω =0.15, then γ' = 6.9 mbar K^{-1} with θ'_b = 49°C (T_b = 39°C) and θ'_a = 31°C(e_a=10 mbar, T_a=30 °C).

Isothermal net radiation R_{ni}

The isothermal net radiation is the net radiation exchange that would occur between the animal's coat surface and the environment if the mean temperature of the coat surface was equal to air temperature T_a. This quantity is defined by

$$R_{ni} = (1-\rho_c)\bar{S} + L_{in} - \sigma T_a^{\,4} \qquad 7)$$

where $L_{in} = \sigma \bar{T}_e^{\,4}$ (W m^{-2}) is the mean thermal irradiance of the animal's body, σ (= 5.67 × 10^{-8} W m^{-2} K^{-4}) is the Stefan-Boltzmann constant, and T_a and $\bar{T}_e$ are expressed in kelvin.

The value of R_{ni} can be calculated from the radiative properties of the coat and climatological records. For example, when T_a = 30 °C, L_{in} = 480 W m^{-2} ($\bar{T}_e$ = 30°C) and $\bar{S}$ = 290 W m^{-2}, then if the reflection coefficient of the coat ρ_c is 0.66 the isothermal net radiation R_{ni} = 100 W m^{-2}.

Rate of change of saturation vapour pressure with temperature Δ_b

The quantity Δ_b is the slope of the straight line drawn on the s.v.p. versus temperature curve to join the points corresponding to the s.v.p. at body-core temperature T_b and the s.v.p. at mean skin temperature $\overline{T}_s$, i.e.

$$\Delta_b = [e_s(T_b) - e_s(\overline{T}_s)]/[T_b - \overline{T}_s] \qquad 8)$$

For example, when T_b = 38.2°C and $\overline{T}_s$ = 35.0 °C then Δ_b = 3.2 mbar K^{-1}, but when the respective temperatures are 41.0 and 39.0°C then Δ_b = 4.0 mbar K^{-1}.

Radiation flux S(z)

In sunshine, the radiation flux $S(z)$ absorbed per unit area by unit depth of coat at a distance z from the skin surface can be calculated using

$$S(z) = \alpha'\overline{p}[S^-(z) + S^+(z)] \qquad 9)$$

The quantities $S^-(z)$ and $S^+(z)$ are the radiation flux densities towards and away from the skin surface respectively, α' is the fraction of the incident energy absorbed by an individual hair, and $\overline{p}$ is the fraction of radiation intercepted by unit depth of coat. The flux densities $S^-(z)$ and $S^+(z)$, and the values of α' and p, can be determined from the solar irradiance of the body and the physical characteristics of the coat respectively, using the expressions derived by Cena & Monteith (1975a). For example, when $\overline{S}$ = 290 W m^{-2}, α' = 0.040, $\overline{p}$ = 18 cm^{-1} and ρ_c = 0.66, then 190 W m^{-2} are reflected to the environment, 25 W m^{-2} are absorbed at the skin surface (reflection coefficient of skin = 0.25) and the remaining 75 W m^{-2} are absorbed within the coat (mean coat depth $\overline{l}$ = 0.5 cm).

Wind penetration depth t

Wind penetration reduces the thermal insulation of an animal's coat. Numerous measurements have been made of the rate of heat transfer through ventilated coats. In the past most measurements have been analysed on the assumption that the decrease in coat resistance is proportional to the square root of windspeed. However, McArthur & Monteith (1980) found that the thermal diffusivity $\overline{\kappa}_{cu}$ (m^2 s^{-1}) of fleece increases linearly with windspeed u (m s^{-1}) according to

$$\overline{\kappa}_{cu} = \overline{\kappa}_{co} + bu \qquad 10)$$

where $\overline{\kappa}_{co} = \overline{l}/\overline{r}_{co}$ is the thermal diffusivity of the coat in still air, and b (m) is a constant. Campbell *et al* (1980) also reported a linear increase with windspeed in the thermal conductance ($1/\overline{r}_c$) of different animal coats. A relationship of the type expressed in equation (10) could arise in several ways, but the simplest postulate is that wind destroys the insulation of a layer of coat, the thickness of which depends on the velocity. We can define the wind penetration depth t by

$$t = bu/[(bu/\bar{l})+(1/\bar{r}_{co})] \tag{11}$$

where the coat resistance in still air $\bar{r}_{co}$ depends on the depth $\bar{l}$ and on the differences of temperature and vapour pressure across the pelage (Cena & Monteith, 1975b,c). For example, when the windspeed is 5 m s^{-1} and b = 0.11×10^{-4}m, then the wind penetration depth $t \simeq 0.5\bar{l}$.

CASE STUDY

Equation (3) will now be used to estimate the heat balance of Jersey cows. Two examples are considered: the influence of windspeed on heat loss in a cold environment; and the effect of solar radiation on heat loss in a hot environment.

Heat loss in a cold environment

Figure 2 shows the dependence on windspeed of the heat flow G_s for a Jersey cow ($\bar{l}$ = 0.5 cm, $b = 0.11 \times 10^{-4}$m) in an environment with an air temperature of 10°C, a vapour pressure of 10 mbar and $R_{ni} = 0$. At low rates of air movement ($u \simeq 0.3$ m s^{-1}) G_s is about 140 W m^{-2}, but the rate of heat loss increases sharply with windspeed to reach about 190 W m^{-2} at u = 5 m s^{-1}. The relation between G_s and u is non-linear: an increase in windspeed

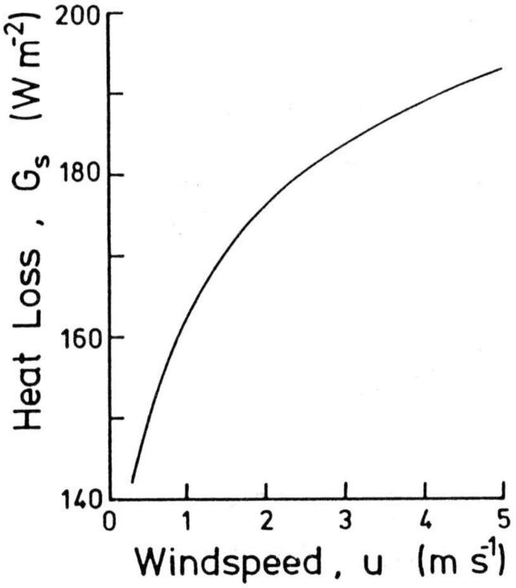

Fig. 2. Heat loss in a cold environment (T_a = 10°C)

from 0.5 to 1.5 m s^{-1} increases G_s by about 20 W m^{-2} whereas between 4 and 5 m s^{-1} the increase is only about 4 W m^{-2}. For an animal with a body mass of 300 kg and a resting metabolic rate less than 150 W m^{-2}, exposure to a windspeed of 5 m s^{-1} can raise its maintenance energy requirement above the 'still air' value by about 20 MJ per day, an increase of more than 30%. Clearly, windspeed is a major determinant of the rate of heat loss from livestock at low air temperature and shelter should minimise (rather than just reduce) the windspeed to which the animals are exposed.

Heat loss in a hot environment

The solid curve (line *a*) in Figure 3 shows the effect of the isothermal net radiation R_{ni} on the heat flow G_s for a cow ($\overline{l}$ = 0.5 cm, thermoneutral heat production = 120 W m^{-2}) in an environment with T_a = 30°C, e_a = 10 mbar and a windspeed of 0.5 m s^{-1}. The value R_{ni} = 0 represents complete shade ($\overline{S}$ = 0, $\overline{T}_c$ = T_a) and R_{ni} = 200 W m^{-2} corresponds to strong sunshine ($\overline{S}$ = 590 W m^{-2}, $\overline{T}_c$ = T_a). The graph indicates that the provision of shade can increase the heat loss G_s from about 90 to 110 W m^{-2}. Of more importance, however, is the corresponding decrease in body-core temperature from 40.9 to 39.3°C (line *b*) and in respiratory rate from about 200 min^{-1} to only 35 min^{-1} (line *c*) if the heat exposure was acute.

In hot climates, forced air movement can relieve thermal strain by increasing the rate of heat loss G_s. Equation (3) can be used to estimate how this loss of heat varies with windspeed for animals in sunlight or in shade.

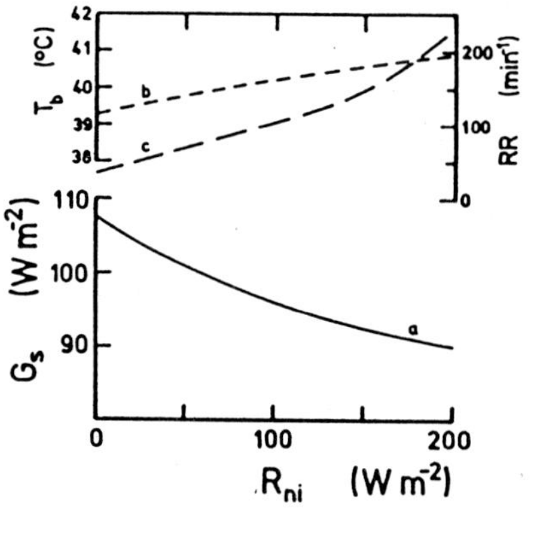

Fig. 3. Heat loss in a hot environment (T_a = 30°C)

SUMMARY

This paper presents an equation to calculate the heat loss and thermal status of animals in relation to different combinations of the climatological variables (temperature, radiation, windspeed and humidity). The equation is developed from heat transfer analysis. If the climatological variables and the relevant heat balance parameters for each class of stock are inserted, the equation can be solved to give the rate at which heat (sensible plus latent) is dissipated from the body surface. Using this equation the relative importance of each variable as a determinant of heat loss, and the need for shelter, can be assessed.

REFERENCES

Bakken, G.S. & Gates, D.M. (1975). Heat-transfer analysis of animals: some implications for field ecology, physiology and evolution. In Perspectives of Biophysical Ecology, D.M. Gates & R.B. Schmerl (Eds.) pp 255-290. New York: Springer Verlag.

Campbell, G.S., McArthur, A.J. & Monteith, J.L. (1980). Windspeed dependence of heat and mass transfer through coats and clothing. Boundary-layer Meteorology, **18**: 485-493.

Cena, K. & Clark, J.A. (1978). Thermal resistance units. Journal of Thermal Biology **3**: 173-174.

Cena, K. & Monteith, J.L. (1975a, b & c). Transfer processes in animal coats. Proceedings of the Royal Society of London B**188**: 377-423.

Finch, V.A. (1972). Thermoregulation and heat balance of the East African eland and hartebeest. American Journal of Physiology **222**: 1374-1379.

Graham, N. McC., Blaxter, K.L., Wainman, F.W. & Armstrong, D.C. (1959). Environmental temperature, energy metabolism and heat regulation in sheep. I. Energy metabolism in closely clipped sheep. Journal of Agricultural Science, Cambridge **52**: 13-24.

Hales, J.R.S. (1974). Physiological responses to heat. In MTP International Review of Science, Physiology Series 1. D. Robertshaw (Ed.) **7**: 107-162. London: Butterworths.

Kelly, C.F., Bond, T.E. & Heitman, H. (1954). The role of thermal radiation in animal ecology. Ecology **35**: 562-569.

McArthur, A.J. (1981). Thermal insulation and heat loss from animals. In Environmental Aspects of Housing for Animal Production. J.A. Clark (Ed.) pp 37-60.

McArthur, A.J. & Monteith, J.L. (1980). Air movement and heat loss from sheep. II. Thermal insulation of fleece in wind. Proceedings of the Royal Society of London B**209**: 209-217.

Monteith, J.L. (1975). Principles of Environmental Physics. London: Arnold.

Monteith, J.L. (1981). Evaporation and surface temperature. Quarterly Journal of the Royal Meteorological Society **107**: 1-27.

Webster, A.J.F. (1974). Heat loss from cattle with particular emphasis on the effects of cold. In Heat Loss from Animals and Man. J.L. Monteith & L.E. Mount (Eds.) pp 205-232. London: Butterworths.

Wiersma, F. & Nelson, G.L. (1967). Nonevaporative convective heat transfer from the surface of a bovine. Transactions of the American Society of Agricultural Engineers **10**: 733-737.

Progress in Biometeorology, 1985, Vol. 2, pp. 59-66

EFFECTS OF PREVIOUS COLD EXPOSURE ON THE COLD RESISTANCE OF YOUNG LAMBS

A.W. Stott
Animal Breeding Research Organisation,
West Mains Road,
Edinburgh, EH9 3JQ.

INTRODUCTION

It has been estimated by Wiener *et al* (1973) that 1.5 to 4 million newborn lambs die annually in Great Britain. Hypothermia is wholly or partly responsible for about half the annual losses excluding stillbirths (Slee, 1976). For this reason mortality figures are extremely weather dependent. Obst & Day (1968) recorded a mortality of 91% in Merino lambs during short periods of heavy wind and rain, which declined to 15% in good weather.

Eales *et al* (1982) have defined two periods of high risk from neonatal hypothermia. The first, from birth to four hours old is largely a problem of excessive heat loss. The animal is saturated in foetal fluid, and metabolic heat production, even if elevated to the normal extent, is often inadequate to maintain homeothermy. The second period occurs between 12 and 36 hours after birth and involves a depressed heat production capability. If the ewe has failed to provide the lamb with sufficient milk to replenish its energy reserves, then heat production will be impaired, and the risk of hypothermia increased. This work suggests that lamb mortality could be reduced by the provision of shelter from birth and for at least 48 hours thereafter. Watson *et al* (1968) significantly reduced mortality by the provision of shelter.

In practice it may be desirable to house the ewes and lambs for longer periods. This may deprive the lamb of the opportunity to acclimatise to cold, resulting subsequently in a reduced ability to resist hypothermia. A balance must be struck for the farm animal between a warm environment that will reduce energy loss, and a variable one which may be expected to produce a more hardy animal (Mount, 1979). Unfortunately there is little information for young lambs, about the development of either cold acclimatisation under natural conditions or controlled acclimation in the laboratory.

In adult sheep acclimation to cold results in a greatly increased resistance to body cooling (Slee & Sykes, 1967). In addition, an experiment on lambs showed a slight, but not significant, increase in cold resistance for lambs housed at 0 °C over those maintained at 25 °C for two weeks after birth (Sykes, 1968).

An important component in the heat production capability of a new born lamb derives from non-shivering thermogenesis (*NST*). The calorigenic response of an animal to noradrenaline (*NA*) can be used to estimate the capacity for *NST* (Jansky, 1973). Brown adipose tissue (*BAT*) is the principal source of *NST* in the newborn lamb; Thompson & Jenkinson (1969) found that the response to noradrenaline declined with age in newborn lambs, and was associated with a similar decline in *BAT*. *NST* accounts for 40% or more of the metabolic response to cold and is of considerable survival advantage (Alexander & Williams, 1968; Samson, 1982). Adult sheep neither respond to noradrenaline nor possess *BAT* (Graham & Christopherson, 1981).

Alexander *et al* (1970) demonstrated that cold acclimation reduced the decline in calorigenic response to *NA* associated with age in newborn lambs. Similar age-related declines in thermoneutral metabolic rate (*TMR*) and peak metabolic response to cold (*PMRC*) were also reduced. The adult sheep does not respond to *NA* following a period of cold acclimation (Graham & Christopherson, 1981) but *TMR* and *PMRC* are thought to rise (Slee, 1974).

This paper examines the importance of cold acclimation on the response of the lamb to hypothermia induced by a standard cold test in a water bath. Possible implications of this for the shelter of newborn lambs are discussed.

MATERIAL AND METHODS

Thirty three Boreray (St. Kilda) Blackface lambs were used in this experiment. These animals were born during May of 1981 or 1982 at the Animal Breeding Research Organisation Field Laboratory, Roslin, Midlothian. Boreray lambs have been found to be more resistant to hypothermia in the field than many commercial breeds of sheep (Slee *et al*, 1980). Their use in this type of experiment may improve future understanding of differences between breeds.

All lambs were suckled naturally before removal from the ewe. They were then weighed and placed in a warm environment (26 °C) for approximately one hour. Measurements of skin thickness were taken at five places on the body (back, mid-side, forequarter, britch and stomach) with Harpenden skinfold calipers. Fleece depths at each site were measured with a probe.

Each lamb was then immersed to the neck in a water bath maintained at 38.5 °C which was approximately thermoneutral (Samson & Slee, 1981). Oxygen consumption was estimated using an open circuit calorimeter incorporating a face mask. Readings were recorded each minute for 15 minutes, and the mean of the lowest 5 consecutive readings was taken as an estimate of thermoneutral metabolic rate (*TMR*). Bath temperature was recorded using two copper-constantan thermocouples 15 cm below the water surface. Rectal temperature was measured with a 6 cm thermocouple probe.

Seven lambs in 1981 were then injected in the mid-back region with 150 μg of noradrenaline per kg body weight. In 1982 the same dose of noradrenaline was infused at 10 μg per kg per min into the jugular vein of a further 7 animals. The metabolic response to this hormone was recorded each minute. The highest mean of 5 consecutive readings was taken as the peak metabolic response (*PMRN*). Recordings continued until metabolic rate returned to pre-injection levels.

A further 19 lambs were subjected to a cooling test in the water bath. Following measurement of *TMR*, the bath was cooled progressively at a controlled rate to give a fall in temperature of about 0.5 °C per minute. Metabolic response and peak metabolic rate (*PMRC*) were measured as previously described. Rectal temperature rose initially and then began to fall steadily. The time taken for the rectal temperature of each lamb to reach 35 °C (from an initial temperature of about 39.5 °C) was recorded as the cold resistance of the animal. At this point the lamb was removed from the water bath and placed in a polystyrene box to recover. After twenty minutes of recovery rectal temperature was recorded, and each lamb was dried under fan heaters and returned to the ewe.

Seven of the 19 lambs which had been cold-tested at birth and 7 of the 14 which had been noradrenaline tested were housed in a climate chamber maintained at 2 °C. The remaining lambs were housed at a temperature of 26 °C (Table 1). The lambs remained at these temperatures with their dams for 14 days. During this period all ewes were fed 0.11 kg of concentrates per head per day and equal quantities of hay in both environments.

After 14 days ewes and lambs were removed from the climate chambers. The lambs were then re-tested, each individual receiving the same type of test (calorigenic response to noradrenaline or water bath cold resistance) as it received at birth.

Full details of the water bath technique are given by Slee *et al* (1980) and Samson & Slee (1981).

Statistical analysis was carried out using the least squares program of Harvey (1972). The data presented here are the fitted means produced by this program. Corrections (by partial regression) have been made where appropriate for liveweight, ambient temperature and deviations of water bath temperature from the standard cooling curve.

Body surface area of the lambs was calculated from liveweight using the equation of Pierce (1934).

RESULTS

Liveweight gain for the experimental period was approximately 2 kg in each environment (Table 1).

Table 1. *Body weight of lambs in each treatment group before and after temperature treatments.*

Number of lambs	Temperature treatment (°C)	Mean body weight (kg ± standard error) At birth (pretreatment)	After 14 days (post-treatment)
14	2	2.7 ± 0.2	4.8 ± 0.2
19	26	2.9 ± 0.2	5.0 ± 0.2

The mean *TMR* of all lambs at birth, was 5.6 ± 0.3 W kg^{-1}. After two weeks at 2 °C or 26 °C *TMR* had declined ($P<0.01$), but treatment effects were not significant (Table 2).

PMRN was 11.5 ± 0.4 W kg^{-1} a birth, 2.1 times *TMR*. This response declined with age ($P<0.05$) but to a lesser extent in cold-treated animals than in warm-treated animals ($P<0.05$) (Table 2). *PMRN* expressed per unit of body surface area did not decline significantly with age, although significant post-treatment differences remained ($P<0.01$) (Table 2).

Mean cold resistance times and recovery temperatures taken at birth were not significantly different from those recorded at two weeks of age. When corrected for liveweight changes these two parameters declined over the experimental period (P 0.01) (Table 3). This occurred despite an increase in fleece depth ($P<0.01$), skin thickness ($P<0.05$) and total body insulation ($P<0.01$) (Table 2).

PMRC fell from 18.7 ± 0.95 W kg^{-1} at birth to 14.4 ± 0.4 W kg^{-1} after two weeks for both treatment groups ($P<0.01$) but was unaffected by age when expressed per unit of body surface area (Table 2).

Table 2. *Thermoneutral metabolic rate* (TMR), *peak metabolic response to cooling in a water bath* (PMRC *and to administration of 150 μg kg⁻¹ noradrenaline* (PMNR). *Data are expressed both on a body weight and a body area basis. Measurements were made on lambs at birth and after two weeks in a cold or warm environment. Standard errors are indicated and significant differences attributable to age of the animal and temperature of the environment are denoted by asterisks in the second and third columns respectively (**,P<0.01; * ,P<0.05).*

	At birth	After 2 weeks at 2 °C	After 2 weeks at 26 °C
TMR (W kg^{-1})	5.6 ± 0.3	4.8 ± 0.3**	4.0 ± 0.3
PMRN (W kg^{-1})	11.5 ± 0.4	8.0 ± 0.6*	5.7 ± 0.6*
PMRN (W m^{-2})	121.4 ± 12.3	121.7 ± 8.6	82.7 ± 12.1**
PMRC (W kg^{-1})	18.7 ± 0.9	14.8 ± 1.2**	13.9 ± 1.2

Total body insulation, fleece depth, skin thickness and *PMRC* measured at two weeks of age were unaffected by previous treatment temperature. Weight-corrected cold resistance time was significantly greater for animals housed at 2 °C (36.3 ± 3.0 min) than for those housed at 26 °C (30.1 ± 2.8 min) (P°0.05) (Table 3).

DISCUSSION

In the adult animal, basal metabolic rate (*BMR*) is proportional to metabolic body size, that is liveweight raised to the power 0.75 (Kleiber, 1961). If this 'surface area law' is related to young growing animals then absolute metabolic rate will increase with age as body size increases, while *BMR* per unit weight will fall. These results and those of Alexander *et al* (1970) confirm this effect for both control and cold acclimated lambs. However, Graham *et al* (1974) suggested that the exponent is higher than 0.75 in young animals. Also Blaxter (1962) has pointed out the frequent deviations of sheep and cattle from the recognised metabolic size

Table 3. *Cold resistance and recovery temperature (R20) for cold-tested lambs. Fleece depth, skin thickness and total body insulation for all lambs corrected and uncorrected for variations in liveweight by partial regression. Data recorded at birth and after 2 weeks in a hot or cold environment. Standard errors denoted, and significant differences attributable to the age of the lamb or the temperature to which it was exposed are indicated by asterisks in the second and third columns respectively (**, P<0.01; *, P±0.05).*

	At birth	After 2 weeks at 2°	After 2 weeks at 26 °C
Cold resistance corrected (min)	48.7 ± 2.9	36.3 ± 3.0**	30.1 ± 2.8*
Cold resistance uncorrected (min)	37.4 ± 1.9	43.0 ± 2.8	38.3 ± 2.2
*R*20 (°C) corrected	36.1 ± 0.4	35.7 ± 0.7**	34.1 ± 0.7
*R*20 (°C) uncorrected	35.7 ± 1.6	40.0 ± 2.6	34.5 ± 2.1
Fleece depth corrected (cm)	1.8 ± 0.06	2.1 ± 0.07**	2.1 ± 0.06
Fleece depth uncorrected (cm)	1.60 ± 0.04	2.30 ± 0.07**	2.20 ± 0.04
Skin thickness corrected (log mm)	141.3 ± 2.9	137.3 ± 2.8	136.0 ± 3.0
Skin thickness uncorrected (log mm)	132.3 ± 2.3	142.0 ± 2.8*	138.5 ± 2.9
Body insulation corrected (°C kg W^{-1})	1.3 ± 0.1	1.4 ± 0.1	1.6 ± 0.1
Body insulation uncorrected (°C kg W^{-1})	1.2 ± 0.1	1.5 ± 0.1**	1.6 ± 0.1

relationships.

The effect of feeding may also cause discrepancies. Eales & Small (1980) reported 3.59 ± 0.01 W kg^{-1} for *BMR* in unfed lambs. The value of *TMR* (fed lambs) given by Alexander *et al* (1970) of 5.7 W kg^{-1} was very similar to that reported here.

PMRN was higher following cold treatment. Alexander *et al* (1970) observed a similar result and related it to higher quantities of *BAT* found in the cold exposed individuals. He also postulated that the normal decline in *BAT* associated with age was completely arrested, as the absolute calorigenic response to noradrenaline found at birth was unchanged following subsequent cold treatment. In this experiment the calorigenic response to noradrenaline was unchanged from birth to two weeks of age in cold acclimated animals, when expressed per unit of body surface area. As surface area increased during the experimental period, *NST* and presumably *BAT* levels must also have risen. This is of survival advantage as heat loss is proportional to surface area (Mount, 1968). Also *NST* can be used preferentially thus avoiding the greater convective losses and impaired mobility associated with shivering.

If *NST* and shivering thermogenesis are additive in their effects, then these factors may have contributed to the improved weight-corrected cold resistance of the cold-treated animals relative to those housed in a warm environment. This seems likely as fleece depth, total body insulation and liveweight did not differ significantly between treatments. Although *PMRC* did not differ significantly between treatments, values for cold-treated animals were slightly higher. The rate of cooling may have been too rapid to discriminate between treatments. Alexander *et al* (1970) was able to show a greater *PMRC* in cold acclimated lambs exposed to wind and low temperatures.

It was found that lambs at two weeks of age were neither more cold resistant nor more able to recover from hypothermia than they had been at birth, despite a rise in total body insulation, greater liveweight and despite no apparent change in *PMRC* expressed per unit surface area. This result was hard to reconcile with that of Eales *et al* (1982) who found that only 9% of the hypothermic lambs in their field study were over 48 hours old. It must be concluded that the newborn lamb outdoors encounters situations of much greater heat loss than older animals. Saturation with foetal fluid, and a high surface area relative to liveweight, would produce such a situation. In addition, the newborn animal cannot carry the energy reserves required to maintain the levels of PMRC measured in the laboratory. In the field this would lead to the starvation/exposure syndrome (Slee, 1977). This theory is supported by Eales *et al* (1982) who reported that only 24% of hypothermic lambs were suffering from excessive heat loss, while 72% had become hypothermic due to lack of energy resulting from starvation. In addition, the newborn lamb may suffer from impaired heat production due to hypoxia during parturition or premature birth. All these factors tend to increase the risk of hypothermia in the young animal.

Despite these reservations, the lamb which receives protection from the elements at birth will rapidly lose the capacity for NST. If this animal is subsequently cold-exposed, its ability to thermoregulate may be impaired relative to other animals that received no shelter.

REFERENCES

Alexander, G. & Williams, D. (1968). Shivering and non-shivering thermogensis during summit metabolism in young lambs. Journal of Physiology **198**: 251-276.

Alexander, G., Bell, A.W. & Williams, D. (1970). Metabolic response of lambs to cold, effects of prolonged treatment with thyroxine and acclimation to low temperatures. Biologia neonatorum **15**: 198-210.

Blaxter, K.L. (1962). The Energy Metabolism of Ruminants. London: Hutschinson Ltd. Eales, F.A. & Small, J. (1980). Summit metabolism in newborn lambs. Research in Veterinary Science **29**: 211-218.

Eales, F.A., Gilmour, J.S., Barlow, R.M. & Small, J. (1982). Causes of hypothermia in 89 lambs. Veterinary Record **110**: 118-210.

Graham, N. McC., Searle, T.W. & Griffiths, D.A. (1974). Basal metabolic rate in lambs and young sheep. Australian Journal of Agricultural Research **25**: 957-971.

Graham, A.D. & Christopherson, R.J. (1981). Effects of adrenaline and noradrenaline on heat production of warm acclimated and cold acclimated sheep. Canadian Journal of Physiology and Pharmacy **59**: 985-993.

Harvey, W.R. (1972). Least squares and maximum likelihood general purpose program. Ohio State University. (Mimeograph).

Jansky, L. (1973). Non-shivering thermogenesis and its thermoregulatory significance. Biological Review, Cambridge Philosophical Society **48**: 85-132.

Kleiber, M. (1961). The Fire of Life: an introduction to animal energetics. London: John Wiley & Sons.

Mount, L.E. (1968). The Climatic Physiology of the Pig. London: Edward Arnold.

Mount, L.E. (1979). Adaption to thermal environment. Man and his productive animals. Contemporary Biology Series, London: Edward Arnold.

Obst, J.M. & Day, H.R. (1968). The effect of inclement weather on mortality of Merino and Corriedale lambs, Kangaroo Island. Australian Society of Animal Production **7**: 239-242.

Peirce, A.W. (1934). Council for Scientific and Industrial Research, Australia Bulletin 84.

Samson, D.E. (1982). Genetic aspects of resistance to hypothermia in relation to neonatal lamb survival. Ph.D. Thesis, University of Edinburgh.

Samson, D.E. & Slee, J. (1981). Factor affecting resistance to induced body cooling in newborn lambs of 10 breeds. Animal Production **33**: 59-65.

Slee, J. (1974). The retention of cold acclimatisation is sheep. Animal Production **19**: 201-210.

Slee, J. (1976). Cold stress and perinatal mortality in lambs. 16th Veterinary Annual pp 66-69. Bristol: Wright

Slee, J. (1977). Cold exposure and survival in new-born lambs. Report to ARC Animal Breeding Research Organisation pp 11-16.

Slee, J., Griffiths, R.G. & Samson, D.E. (1980). Hypothermia in newborn lambs induced by experimental immersion in a water bath and by natural exposure outdoors. Research in Veterinary Science **28**: 275-289.

Slee, J. & Sykes, A.R. (1967). Acclimatisation of Scottish Blackface sheep to cold 1. Rectal temperature responses. Animal Production **9**: 333-347.

Sykes, A.R. (1968). A study of the variation in response to climatic stress within and between breeds of sheep. Ph.D. Thesis, University of Edinburgh.

Thompson, G.E. & Jenkinson, D. McE. (1969). Non-shivering thermogenesis in the newborn lamb. Canadian Journal of Physiology and Pharmacy **47**: 249-253.

Watson, R.H., Alexander, G., Cumming, I.A., MacDonald, J.W., McLauglin, J.W., Rizzoli, D.J. & Williams, D. (1968). Reduction of perinatal loss of lambs in winter in Western Victoria by lambing in sheltered individual pens. Proceedings of the Australian Society of Animal Production **7**: 243-249.

Wiener, G. Deeble, F.K., Broadbent, J.S. & Talbot, M. (1973). Breed variation in lambing performance and lamb mortality in commercial sheep flocks. Animal Production **17**: 229-243.

Progress in Biometeorology, 1985, Vol. 2, pp. 67-82

SHELTER FOR ANIMALS IN HOT COUNTRIES

Ruth M. Gatenby
University of Edinburgh,
Centre for Tropical Veterinary Medicine,
Easter Bush,
Roslin, Midlothian,
Scotland, EH25 9RG.

Shelters or houses are used to improve the productivity of domestic animals in hot countries. They may be constructed for many reasons: to protect animals from cold, to reduce the disease hazard, to change the lighting pattern or to make the management of the animals easier. Although a shelter may be constructed with only one of these aims in mind, every aspect of the environment will be altered by the presence of the shelter. For instance, laying hens may be kept in a large house containing battery cages because this eases management. But keeping high densities of birds is unlikely to be successful unless the design of the house ensures that the birds are kept cool by good shading and adequate ventilation. Animals confined within a shelter are unable to move to another environment if that within the shelter is unsuitable, so it is important that an appropriate environment is provided.

In this paper the design of shelters with respect to the thermal environment is considered in detail and other aspects considered briefly. The physical principles apply to all species of domestic animals and humans.

HEAT BUDGET

An animal standing in the sun has two important inputs of heat: it absorbs short-wave radiation and it has an internal metabolic heat production. The animal loses heat as sensible heat (convection, long-wave radiation and conduction) and by evaporation of water. In a hot environment the animal suffers from heat stress because its heat inputs are large compared with its heat losses. An animal suffering from acute heat stress gets too hot and may even die. However, chronic heat stress is more important economically because in the long term, a heat stressed animal has a low voluntary food intake and is unproductive.

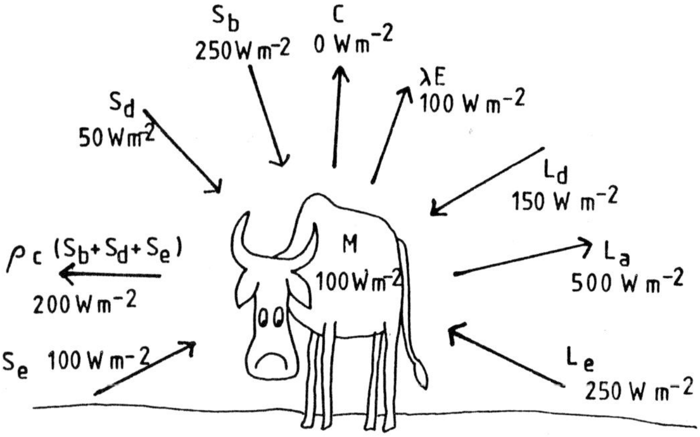

Figure 1. The heat budget of an animal standing in the sun, showing typical values of the components. M is the metabolic heat production, S_b is the incident short-wave radiation in the direct beam, S_d is the incident diffuse short-wave radiation from the sky, S_e is the short-wave radiation received after reflection from the environment, ρ_c, $(S_b+S_d+S_e)$ is the short-wave radiation reflected away from the animal, L_d is the long-wave radiation received from the sky, L_e is the long-wave radiation received from the environment, L_a is the long-wave radiation emitted by the animal, C is the heat loss by convection, and λE is the heat loss by evaporation. All fluxes are expressed per unit of body surface area.

The main components of the heat budget of an animal standing in the sun are shown in Figure 1. Typical values of the fluxes for an ox are shown, assuming that air temperature and the average temperature of the animal's coat are both about 35° C. The net gain of short-wave radiation is 250+50+100–200 = 200 W m^{-2}, and about half of this heat is lost as long-wave radiation (net long-wave loss = 500–150–250 = 100 W m^{-2}). Metabolic heat production adds a further 100 W m^{-2} to the heat budget, so that the sum of the net radiative gain plus metabolism is 200–100+100 = 200 W m^{-2}.

In this simplified case there is practically no net heat transfer by convection because air temperature equals the mean temperature of the coat surface (but in practice there is convective heat loss from the parts of the body in the sun, and advection to the lower parts of the body). The net gain of heat by radiation and metabolism must therefore be lost by the

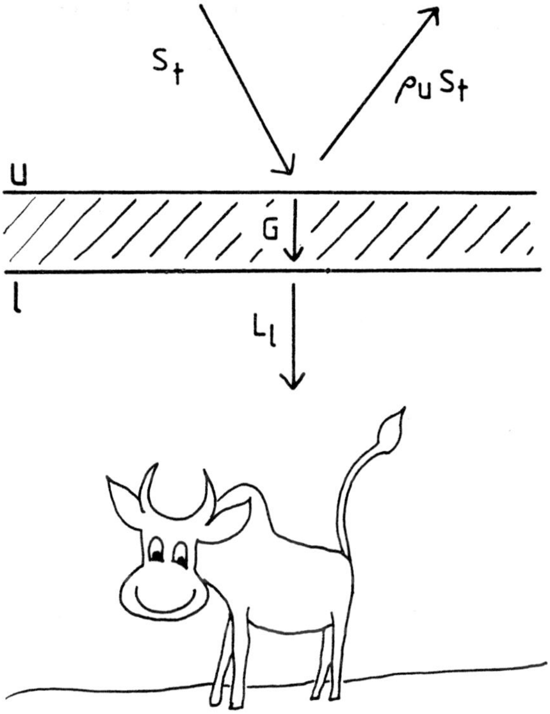

Figure 2. The roof: u and l are the upper and lower surfaces respectively, S_t is the total short-wave radiation incident on u, ρ_u is the reflection coefficient of u, G is the heat transfer down through the roof, and L_l is the long-wave radiation emitted by l.

evaporation of water or else be stored in the body. Typically the evaporative heat loss from an ox in the sun will be 100 W m^{-2} so that the remaining 100 W m^{-2} are stored in the body.

The aim of shelter to reduce heat stress is therefore to reduce the absorption of short-wave radiation, and to increase the heat losses by convection, long-wave radiation, conduction and evaporation.

PRINCIPLES OF THE DESIGN OF SHELTERS

Roof

The roof protects the animal from direct short-wave radiation. In addition it reduces the heat gained by diffuse radiation from the sky and by short-wave radiation reflected from the ground. Measurements of the radiation received by shaded and unshaded animals in the southern U.S.A. are given by Bond *et al* (1967). Poor shade design may reduce the heat lost as long-wave radiation and by reducing air movement, reduce heat losses by convection and evaporation. Even in hot countries, animals may suffer from cold stress under certain circumstances. By protecting animals from rain and by reducing long-wave heat losses at night, shelters can help to prevent cold stress.

The shelter must provide shade for all the animals throughout the day. In other words the shadow of the roof must always fall within the area in which the animals are confined. The radiation budget of the roof is shown in Figure 2. The temperature (T_u) of the upper surface of the roof gets very high in direct sunlight and in California T_u may be 55° C higher than air temperature (Neubauer & Cramer, 1965). L_l, the long-wave radiation emitted by the lower surface (*l*) of the roof is one of the major heat loads on the animals under the roof (Kelly *et al*, 1954). L_l is strongly dependent on the temperature, T_l, of *l* so that in a well designed roof, T_l is kept as low as possible. There are several ways in which this can be done:

(i) by building the roof from a material which is a good insulator of heat so that the rate of heat transfer (*G*) down through the roof from u to l is low. The insulation of a roof is the product of its insulation per unit depth and its thickness. The insulation values of some roofing materials are shown in Table 1. Asbestos, plaster and wood are almost 1000 times better insulators than metal roofing sheets, and cellular polystyrene and felt are even better. Thatch too, is a good insulator. A concrete roof keeps the environment within a house cooler than a thin aluminium roof (Ganamurthy *et al*, 1980), and a thick layer of thatch is ideal. For a thin metal roof, T_l can be considerably reduced by putting a layer of insulating material under the original roof (Figure 3a). In South-East Asia animals are stalled under human dwellings (Figure 3b), this providing a well-insulated roof.

Table 1 *Thermal insulation of roofing materials (adapted from Kaye & Laby, 1978; Wathes, 1981).*

Material	Insulation per unit depth ($K\,m\,W^{-1}$)
Cellular polystyrene	34
Felt	25
Plaster	8
Wood	7
Asbestos cement sheet	2.5
Concrete	0.7
Metal roofing sheets	0.01

Figure 3. Animal houses. A, The use of reeds under corrugated metal roof in the Sudan; B, Cattle under a house in South-East Asia; C, Open walls to allow movement of air over battery hens; D, Sheep house with open walls and temporary protection against storms.

(ii) by increasing the reflection coefficient (r_u) of the upper surface of the roof. Light coloured and shiny roofs have higher values of r_u than dark roofs, so that painting a rusty roof with white paint is beneficial.
(iii) by sprinkling the roof with water (assuming it is available) to lower T_u by evaporative cooling.
(iv) by constructing the roof from materials which allow the movement of air through them so that heat is more readily lost by free convection. Both thatch and tiles allow this movement of air.

The higher the roof the more the animal is exposed to the cool sky (Garrett *et al*, 1967) and the more ventilation can take place under the roof.

Some authors (e.g. Kelly *et al*, 1965) recommend that the roof should be large so that a large area of ground around the animals is protected from direct solar radiation, thus reducing S_c and L_c. The reflection coefficient of the lower surface of the roof has little effect on its heat budget.

A large overhang of the roof at least 1 metre outside the walls is desirable to ensure that the animals inside the shelter are always protected from direct sunlight and from driving rain. However a large overhang makes the roof more susceptible to damage by wind (Eaton, 1980) and it may need to be supported at the edge. The pitch of the roof also affects its stability in strong winds. The steeper the roof, the more stable it is, the less likely it is to leak, and the cooler the environment within the house.

The roof is the most important part of a shelter for an animal in hot countries, and even cheap structures can provide good protection against thermal stress if well-designed. Typically the black globe temperature in a hot environment may be reduced by 10 °C by shading (Givens, 1965).

Walls

Three types of wall design are used in shelters in hot countries. The most common is the shelter with open sides; either the sides are completely open or there is a solid wall of not more than 1 metre high and thereafter wire netting. The other two types of walls are the traditional thick walls seen in human houses in hot dry regions, and modern animal houses with complete walls and controlled ventilation in which the idea is to have a completely controlled environment.

Air movement through the shelter keeps the air within the shelter similar to that outside, removing sensible heat, moisture, and pollutants from the animals and the floor. In hot countries a high ventilation rate is usually desirable to keep the animals cool and dry. Most builidings rely on natural convection; i.e. the air moves of its own accord, and the shelter design must ensure that the ventilation rate is adequate in all weather. It must be sufficient in calm weather, but animals must have enough protection in a hurricane. In the absence of wind, the air inside the house moves in response

Figure 4. Animal houses. A, Traditional thick-walled house for scavenger poultry in northern Nigeria; B, Ducks housed above a fish pond so that their droppings fall into the pond; C, Fresian cow standing under a water sprinkler to increase heat loss by evaporation; D, Traditional shelter for calves in the Sudan, made from local materials.

to buoyancy. Hot, moist air rises from the animals, taking heat and water vapour with it. This free convection is encouraged if the walls are open, if the roof has a steep pitch and the ridge is open or if the roof itself is permeable to air. The movement of air in response to wind is called forced convection and results in the transfer of heat and water vapour laterally out of the shelter. For birds in stacks of battery cages a lateral flow of air is desirable to cool the upper tiers which, in a house in which the ventilation is dominated by natural convection, are much warmer that the cages near the ground. Figure 3c shows battery hens in a naturally ventilated building with open sides.

Because natural ventilation is cheap and generally reliable, most animal houses in the tropics rely on it. They are designed with open sides, with possibly some protection for susceptible animals aginst strong winds and rain either by a solid structure on one or more walls, or by temporary protection being provided in certain seasons of the year (Figure 3d).

The second type of animal house found in the tropics has thick walls. Figure 4a shows a small thick walled house for scavenging poultry. These walls have a large thermal capacity so that they absorb rather than transmit the incident short-wave radiation during the day and release it at night. Inside a thick walled building there is little diurnal variation in temperature, and the maximum temperature occurs several hours after the maximum external temperature. For instance, in an environment with a maximum air temperature of 40 °C at 2 p.m. and a minimum air temperature of 10 °C in the early morning, the temperature inside an empty building with walls 0.5 m thick will reach a maximum at about 5 a.m., 15 hours after the external maximum (Curd, 1976), but the diurnal variation in internal temperature will be small, tending to a mean of 25 °C. The ventilation rate of a thick walled house is low during the day to reduce the rate at which it heats up, although the doors may be opened at night to enhance heat loss. The animals lose a large proportion of their heat by radiative exchange with the cool walls.

Thick walled houses are found only in the arid and semi-arid areas of the tropics where there is a large diurnal variation in temperature. In the humid tropics, not only is there little diurnal variation in ambient temperature, but also a high ventilation rate is required in animal houses to remove moisture. Even in the dry tropics this type of house is used only for small numbers of animals, otherwise the low ventilation rate cannot remove enough water vapour. Thick walled buildings are particularly desirable where air temperature exceeds body temperature during the day, because under these conditions a high ventilation rate aggravates sensible heat stress even though it allows a rapid evaporation rate.

The third type of shelter is the controlled environment house. This is totally enclosed and relies on fans for ventilation. In theory the capacity of fans needed in any set of circumstances can be calculated (Sainsbury & Sainsbury, 1979), but in practice the data needed for these calculations are not known accurately. The system must be able to cope with the removal of heat and water vapour under all conditions including very hot days and very windy days when the fans will not work to capacity. Controlled environment houses are very expensive, and because of their dependence on a reliable supply of electricity and maintenance services they are not recommended for general use in less developed hot countries, but only when an enclosed house is required for other reasons such as to give control over the lighting pattern.

Floor

The floor of an animal house has comparatively little effect on the thermal environment. Slatted floors are expected to encourage air movement in the vicinity of the animals, but in practice the type of floor (unless it is refrigerated) has little effect on animal performance if evaporative cooling is available (Morrison *et al*, 1979). The design of the floor is more important in its effect on the disease environment. Unless the faeces and urine are disposed of properly, the environment becomes moist and ideal for the growth of pathogenic micro-organisms. Good drainage is essential. There are basically three ways in which manure is dealt with: by frequent removal, by allowing a build-up usually in bedding (the deep litter system) and by slats. Figure 4b shows ducks in a house above a fish pond. Their droppings fall through the mesh into the pond below.

Siting

The siting of an animal house affects the environment inside it. Ferguson (1970) discussed the ways in which the trees and buildings around the animal house affect the airstream around and through the building. Tall trees growing next to the house shade the roof and thus help to keep it cool. The nature of the ground surrounding the house affects the reflection of long-wave radiation up into the house and the temperature of the airstream approaching the house, so that it is desirable to surround the house by short vegetation rather than concrete or bare earth.

Artificial Cooling

As well as designing animal houses to reduce heat stress, artificial cooling may be provided. The evaporation from animals can be increased by wetting them, air in the house can be cooled to increase the potential for convective heat loss, and air movement enhances convection and evaporation. The most cost effective methods of cooling are those which cool the animal directly, rather than attempting to cool the environment within the shelter.

Evaporation is a good method of cooling in hot environments because the rate of evaporation depends not on air temperature but on the absolute humidity of the air. In a dry environment the evaporation rate from an animal is limited by the rate at which it sweats and pants, so that the rate of evaporation can be enhanced by wetting the body surface. This can be achieved simply be splashing water over the animal (Sinha & Minett, 1947), but in commercial units the most common way of applying water is to provide showers (Miller *et al*, 1951; Ansell, 1976; Thiagarajan *et al*, 1978). It is not necessary for the shower to be heavy or continuous. Most cooling results from the evaporation of the water not from conduction between the body and the cooler water, so that once the animal is wetted, the evaporative cooling continues for a period of about one hour (Rafai & Papp, 1976) depending on ambient temperature and humidity and how much water is held in the pelage. Hsia *et al*, (1974) found that sprinkling fattening pigs with only 50 cm^3 of water per pig every 45 minutes increased liveweight gains. Sprinkling every 90 minutes was less satisfactory. Because an animal knows when it needs to be cooled, it is possible to have the shower in only one area of the shelter and allow the animals to stand under the shower when they want to (Seath & Miller, 1948). The size of the shower must be large enough to allow even subordinate animals to get wet and not be pushed out by dominant animals.

Showers and sprays are used for evaporative cooling of cattle (Figure 4c), buffaloes and pigs. Wallows also serve a similar function for buffaloes (Minett, 1947) and pigs (Rannfelt & Kroeske, 1974), species which naturally wallow. Wallows for pigs require about 5 litres of water per pig per day (Rannfelt & Kroeske, 1974), less than is conventionally used in showers. Muddy wallows are beneficial from the cooling point of view - a layer of mud takes longer to dry than a layer of water - but may not be recommended for hygiene reasons.

The air within the house can be cooled by the evaporation of water, provided that the climate is reasonably dry. The most effective way of doing this is by using "foggers" to give a fine mist which evaporates near the animals and cools the air (Wilson *et al*, 1957). Wet pads on the wall have a similar purpose but are less effective, and water lying on the floor is sometimes said to be there to cool the air by evaporation! For enclosed buildings air may be passed through refrigeration units and recycled. The design of refrigeration units is discussed by Ogston (1976). The cost of sophisticated refrigeration systems is too high for their use in commercial animal production, but simple evaporative cooling of air is used in poultry houses where wetting individual birds is impracticable.

The heat loss by conduction to the floor from finishing pigs in a hot environment can be over 300 W m^{-2} per unit area of contact (Spillman & Hinkle, 1971) or almost 100 W m^{-2} averaged over the whole body surface. But in a hot environment, heat loss to the floor is not enhanced by lowering the temperature of the floor (Kelly *et al*, 1964) so that refrigeration of the floor is not recommended. The provision of cool drinking water is beneficial because animals lose heat by warming it up to body temperature. If the water is initially at 10 °C the heat lost in this way by non-lactating cattle is equivalent to a continuous loss of about 15 W m^{-2}, and is even more for high yielding dairy cows which consume vast quantities of drinking water.

Fans within a house increase the rate of air movement and thus reduce the boundary layer resistance around the animals. Fans have little effect on heat loss by convection in hot environments because there is little difference in temperature between the animal's surface and the air. In houses for humans in hot countries fans are common because they increase the rate of evaporation of sweat so that people feel more comfortable and cooler. However, non-working animals rarely become covered in sweat, so that the rate of evaporation is almost independent of the boundary layer resistance, and although some authors (Thiagarajan *et al*, 1977) do recommend the use of fans, other workers have found that fans are relatively ineffective in cooling animals if evaporative cooling is not provided (Garrett *et al*, 1960).

PRACTICALITIES OF THE DESIGN OF SHELTERS

In practice, animal houses are not built solely to attain the optimum thermal environment. Consideration must also be give to other attributes of the house such as ease of feeding and observing the animals and to economics. Building shelters is expensive in terms of both materials and labour. In developed countries it is possible to quantify the economics of housing (Hoglund & Albright, 1970), but in less developed countries this is much more difficult. Nevertheless, some general points can be made. The design for optimum economic performance will generally be less sophisticated and give less environmental modificaion that the design which gives the optimum biological performance. Thus, although it is widely accepted that evaporative cooling increases the performance of dairy cows in the southern U.S.A., only for high yielding dairy herds in the hottest areas are the benefits sufficient for evaporative cooling to be recommended (Brown *et al*, 1974).

For extensive animal production systems any shelters must be very simple if they are to be economical. For extensive production of ruminants, for example, apart from pens in which the animals are enclosed at night, trees are the only economical form of shelter. For semi-intensive systems, such as dairying (Parks & Skinner, 1957; MacFarlane & Stevens, 1972), beef finishing (Kelly *et al*, 1965) or lamb finishing (Pritchard & Ruxton, 1977), it may be worthwhile to provide artificial shades (Figure 4d) and possibly showers, but no other form of cooling. Animals in intensive production systems, such as modern egg producing systems, are usually housed in open sided buildings. Completely controlled environment houses can give dramatic increases in animal productivity (Thatcher *et al*, 1974) but are rarely justified economically.

MacFarlane (1981) pointed out that traditional animal shelters have grown out of the needs, resources and ingenuity of farmers. The costs of local materials and traditional skills are much lower than the costs of imported materials and technologies, so that traditional types of animal shelter can often by justified economically where "modern" structures are prohibitively expensive. In most parts of the tropics skilled craftsmen use straw or other materials to make thatched roofs (Hall, 1981). However, traditional shelters constructed from wood, mud and thatch harbour vermin, are readily destroyed by fire and may be considerably less durable than modern constructions so that modern structures are preferred in large animal production units.

CONCLUSION

Shelters for animals in hot climates may be built primarily to reduce heat stress on animals, but even if they are built for other reasons, the thermal environment inside the shelter has a large effect on the productivity of the animals within it. The roof should be thick, reflective and high, and the shade it provides must always be accessible to the animals. Open walls are often satisfactory, and the floor should be well-drained. Evaporative cooling using showers or wallows is the most common form of artificial cooling.

REFERENCES

Ansell, R.H. (1976). Maintaining European dairy cattle in the Near East. World Animal Review, **20**: 1-7.

Bond, T.E., Kelly, C.F., Morrison, S.R. & Pereira, N. (1967). Solar, atmospheric and terrestrial radiation received by shaded and unshaded animals. Transactions of the American Society of Agricultural Engineering, **10**: 622-625,627.

Brown, W.H., Ruquay, J.W., McGee, W.H. & Iyengar, S.S. (1974). Evaporative cooling for Mississippi dairy cows. Transactions of the American Society of Agricultural Engineering, **17**: 513-515.

Curd, E.F. (1976). Heat losses and heat gains. In Control of the Animal

House Environment, T. McSheey (Ed.) pp 153-183. London: Laboratory Animals Ltd.

Eaton, K.J. (1980). Making buildings to withstand strong winds. Appropriate Technology **7**(3): 21-23.

Ferguson, W. (1970). Poultry housing in the tropics: applying the principles of thermal exchange. Tropical Animal Health and Production, **2**: 44-58.

Ganamurthy, V., Shanmugasundaram, S., Viswanathan, R.S. & Thiagarajan, M. (1980). The study of environments under two roofing conditions on factors related to some production traits in White Leghorns. Cheiron, Tamil Nadu Journal of Veterinary Science and Animal Husbandry, **9**: 82-87.

Garrett, W.N., Bond, T.E. & Kelly, C.F. (1960). Effect of air velocity ' on gains and physiological adjustments of Hereford steers in a high temperature environment. Journal of Animal Science, **19**: 60-66.

Garrett, W.M., Bond, T.E. & Pereira, N. (1967). Influence of shade height on physiological responses of cattle during hot weather. Transactions of The American Society of Agricultural Engineering, **10**: 433-434,438.

Givens, R.L. (1965). Height of artificial shades for cattle in the southeast. Transactions of the American Society of Agricultural Engineering, **8**: 312-313.

Hall, N. (1981). Has thatch a future? Appropriate Technology **8**(3): 7-9

Hoglund, C.R. & Albright, J.L. (1970). Economics of housing dairy cattle. A review. Journal of Dairy Science, **53**: 1549-1559.

Hsia, L.C., Fuller, M.F. & Koh, F.k. (1974). The effect of water sprinkling on the performance of growing and finishing pigs during hot weather. Tropical Animal Health and Production, **6**: 183-187.

Kaye, G.W.C. & Laby, T.G. (1978). Tables of Physical and Chemical Constants, 14th edition. London: Longman.

Kelly, C.F., Bond, T.E. & Garrett, W. (1964). Heat transfer from swine to a cold slab. Transactions of the American Society of Agricultural Engineering, **7**: 34-35,37.

Kelly, C.F., Bond, T.E. & Heitman, H. (1954). The role of thermal radiation in animal ecology. Ecology **35**: 562-569.

Kelly, C.F., Curley, R.G. & Allen, W.S. (1965). Beef production (in hot climates). Agricultural Engineering, **46**: 80-81.

MacFarlane, J.S. & Stevens, B.A. (1972). The effect of natural shade and spraying with water on the productivity of dairy cows in the tropics. Tropical Animal Health and Production, **4**: 249-253.

MacFarlane, W.V. (1981). The housing of large mammals in hot environments. In Environmental Aspects of Housing for Animal Production, J.A. Clark (Ed.) pp 259-284. London: Butterworths.

Miller, G.D., Frye, J.B., Burch, B.J., Henderson, P.J. & Rusoff, L.L.

(1951). The effect of sprinkling on the respiration rate, body temperature, grazing performance and milk production of dairy cattle. Journal of Animal Science, **10**: 961-968.

Minnet, F.C. (1947). Effects of artificial showers, natural rain and wallowing on the body temperature of animals. Journal of Animal Science, **6**: 35-49.

Morrison, S.R., Heitman, H. & Givens, R.L. (1979). Effect of air movement and type of slotted floor on sprinkled pigs. Tropical Agriculture, Trinidad, **56**: 257-258.

Neubauer, L.W. & Cramer, R.D. (1965). Shading devices to limit solar heat gain but increase cold sky radiation. Transactions of the American Society of Agricultural Engineering, **8**: 470-472,475.

Ogston, W.M. (1976). Refrigeration. In Control of the Animal House Environment, T. McSheehy (Ed.) pp 129-152. London: Laboratory Animals Ltd.

Parks, R.R. & Skinner, T.C. (1965). Dairy production in hot climates. Agricultural Engineering, **46**: 78-79.

Pritchard, C.J.R. & Ruxton, I.B. (1977). A preliminary investigation into the effect of shade on the growth and feed intake of weaned Awassi ewe lambs. Joint Agricultural Research and Development Project Publication No. 79. Bangor: University College of North Wales, and Saudi Arabia: Ministry of Agriculture and Water.

Rafai, P. & Papp, Z. (1976). Effect of the moistening of body surface on heat production and heat sensation in fattening pigs. Acta Veterinaria Academiae Scientiarum Hungaricae **26**: 95-104.

Rannfelt, C.A. & Kroeske, D. (1974). Pig housing in warm climates. World Animal Review, **10**: 24-30.

Sainsbury, D. & Sainsbury, P. (1979). Livestock Health and Housing. London: Balliere Tindall.

Seath, D.M. & Miller, G.D. (1948). A self-serving sprinkling device for cooling dairy cattle. Journal of Animal Science, **7**: 251-256.

Sinha, K.C. & Minett, F.C. (1947). Application of water to the body surface of water buffaloes and its effect on milk yield. Journal of Animal Science, **6**: 258-264.

Spillman, C.K. & Hinkle, C.N. (1971). Conduction heat transfer from swine to controlled temperature floors. Transactions of the American Society of Agricultural Engineering, **14**: 301-303.

Thatcher, W.W., Gwazdauskas, F.C., Wilcox, C.J., Toms, J., Head, H.H., Buffington, D.E. & Fredriksson, W.B. (1974). Milking performance and reproductive efficiency of dairy cows in an environmentally controlled structure. Journal of Dairy Science, **57**: 304-307.

Thiagarajan, M., Michael, R.D. & Prabaharan, R. (1978). Effect of sprinkling water on cross-bred cows during summer. Kerala Journal of Veterinary Science, **9**: 215-220.

Thiagarajan, M., Shanmugasundaram, S. & Michael, R.D. (1977). Effects of increased air velocity on the comfort and production in layers during summer. Cheiron, Tamil Nadu Journal of Veterinary Science and Animal Husbandry, **6**: 164-169.

Wathes, C.M. (1981). Insulation of animal houses. In Environmental Aspects of Housing for Animal Production, J.A. Clark (Ed.) pp 379-412. Lon don: Butterworths.

Wilson, W.O., Hart, S.A. & Woodward, A.E. (1957). Mist cooling hens in cages by fogging. Poultry Science, **36**: 606-613.

Progress in Biometeorology, 1985, Vol. 2, pp. 83-98

SHELTER STUDIES USING THERMAL MODELS OF CATTLE

C.G. Jones and J.M. Bruce
SFBIU,
Craibstone,
Bucksburn,
Aberdeen,
AB2 9TR.

INTRODUCTION

The measurement of climate modification by shelters and the specification of thermal requirements of animals within these shelters has been the subject of study within many scientific disciplines. Compiling these studies into a system for application to practical agriculture requires validation through experiment. The objective of this research is to study the interactions of climate, shelters and cattle using physical and mathematical models of the system. The aim of this approach is to simulate the main elements of the system and to provide a practical solution with the appropriate accuracy.

SHELTER/CLIMATE INTERACTIONS

Cattle shelters in the United Kingdom are usually simple uninsulated structures relying on natural ventilation to produce the internal thermal environment. The requirements for thermal design specifications range from determining orientation, height and porosity of a windbreak to quantifying sizes and positions of ventilation inlets/outlets, and specifying orientation for a fully roofed and walled structure. Climate modification by shelters within this range has generally been studied by separating the climate into its major components and applying results from theory and experiment to interactions between the shelter and each component. Examples of this are studies of wind reduction by windbreaks (Perera, 1981), shading by roofed shelters (Bond *et al*, 1976), and ventilation of open fronted (Koenig, 1978) and enclosed (Bruce, 1978) cattle buildings. The complex interactions of these components have been studied using theoretical (Albright, 1973) and physical (Koenig, 1978) models of buildings. However, such models are usually applicable to an enclosed type of structure, are rarely tested using full scale buildings and do not include the influences of the resulting internal thermal environment on the enclosed animals.

ANIMAL/CLIMATE INTERACTIONS

Heat losses from an animal to its surroundings form part of its energy balance which can be described by

$$F_i = E_m + E_g + J + Q_s + Q_c \qquad 1)$$

where the symbols are defined in the appendix.

When heat losses from an animal are in excess of heat produced due to maintainance and the inefficiencies of production (thermoneutral heat production), then metabolisable energy is redirected from production (e.g. liveweight gain) towards heat production for maintainance of homeothermy.

Experiments comparing live-weight gains of cattle in different shelters have given conflicting results because of differences in breed, age, shelter type and climate (Jordan *et al*, 1969; McCarrick & Drennan, 1972 a,b; Jorgensen *et al*, 1970). These factors limit the scope for future application of this type of research. It is more appropriate to study the relationships between the climate and heat loss from animals, and to incorporate the results into a model of the animal's energy balance for prediction of liveweight changes.

The climate can be described by four major variables: air temperature, windspeed, net radiation and rainfall. These combine to form a potential for net radiative, convective and conductive heat loss from the animal; defined as climatic energy demand (sensible heat loss). Evaporative heat losses are minimal and fairly constant in cold conditions with a value of approximately 17 W m^{-2} (Blaxter & Wainman, 1961; Gonzales-Jiminez & Blaxter, 1962).

Experiments on cattle in controlled environment chambers have resulted in values for heat production and thermal resistances over a range of air temperatures (–20 °C to +20 °C) but at low windspeeds, and under balanced net radiative conditions (Blaxter & Wainman, 1960; Webster *et al*, 1970). Some studies have included varying windspeeds (0.2 to 5.3 m s^{-1}) (Blaxter & Wainman, 1964; Webster *et al*, 1970) and rainfall (25 mm hr^{-1}) (Holmes & McLean, 1975). The complexity of simulating natural climatic conditions within a chamber means that measurements and predictions of heat losses need to be made outside.

Thermal models of animals are simpler alternatives to using live animals outside. These simulate the thermal characteristics of the animal (derived from laboratory studies) and can be used to develop and validate predictive equations for heat losses. For example, Wiersma (1967) measured convective heat losses from a hide and hair covered model of a steer in a wind tunnel; Webster (1971) measured heat losses from a hairless model of a steer in different shelter regimes outside, and McArthur & Monteith (1980) studied convective heat losses from a model sheep in a wind tunnel and outside.

The research described here uses thermal models of cattle to investigate the shelter provided in winter by simple structures and to provide data for development and validation of a mathematical model for prediction of heat losses from cattle under each shelter regime.

DEVELOPMENT AND VALIDATION OF A PREDICTIVE MODEL

Figure 1 shows the design for a thermal model of a 500 kg suckler cow with simulated tissue insulation of 0.2 ° C m^2W^{-1} (Burnett & Bruce, 1978), and covered with a cured hide.

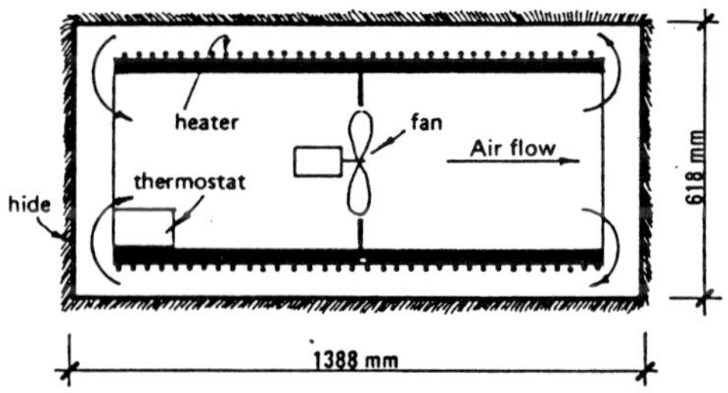

Fig 1. Thermal model of a suckler cow (Bruce, 1980)

The model was placed on an exposed experimental site for four consecutive winters (September – April), 1978 – 1982. The electrical energy required to maintain internal temperature at 39 °C was recorded automatically every hour together with measurements of air temperature, net radiation (over grass), rainfall and windspeed. The data from 1978-1979 was used for developing the predictive model shown in equation (2) (Bruce, 1980).

$$Q_s = \frac{T_b - T_a - \alpha I_a R}{I_t + I_h + I_a} \quad 2)$$

The parameter α mainly describes the radiation geometry of the model.

I_a is calculated from:

$$I_a = (5.3 + 7u^{0.6})^{-1} \quad 3)$$

where the value 5.3 describes long wave radiative exchange between the coat surface and the surroundings (Bruce, 1980).

I_h is calculated from:

$$I_h = \beta(1 - \min(\gamma v, \delta)) \qquad 4)$$

where β is the maximum value and is reduced by the rain term described by

$$v = r(t) + v(t-N)e^{-aN} \qquad 5)$$

(Bruce, 1980). The number of hours since last rainfall N is zero both for $N>24$ and for prediction of daily mean Q_s values.

The parameters α, β, γ, δ and a were optimised for hourly and daily mean measurements to give the minimum error sum of squares between observed and predicted Q_s values.

For hourly predictions

$$\alpha = 0.62, \beta = 0.066, \gamma = 0.3, \delta = 0.6, a = 0.2$$

For daily predictions

$$\alpha = 0.75, \beta = 0.068, \gamma = 1.5, \delta = 0.55$$

Comparison of hourly and daily predictions and measurements of Q_s gave comparable errors over the three following winters. Coefficients of variation were approximately 6% and 4% for hourly and daily predictions respectively (Bruce, 1980).

A second thermal model was placed in a fully enclosed and occupied cattle building for the winters between 1979 and 1982. Measurements of daily mean Q_s gave a reduction of unsheltered Q_s values of 13% in autumn and spring and 20% in mid-winter (Bruce, 1982). A model developed by Webster (1971) predicted daily mean heat losses from a thermal model (without a hair coat) with a coefficient of variation of 8% for extreme air temperatures, low windspeeds and zero rain. Measurements of heat losses from the model when placed in an open fronted shed were between 78% and 81% of measurements in an exposed paddock (Webster *et al*, 1970). This is surprisingly close to our measurements considering the differences between climate, models and buildings.

The predictive accuracy of the mathematical model was further tested by including estimations of Q_s in an energy balance model for suckler cows at three feed levels (Bruce, 1980). Figure 2 shows predicted liveweight changes for unsheltered suckler cows over a 29 week winter. The agreement with mean liveweight measurements for 24 cows on a nearby unsheltered site is sufficiently good to indicate the usefulness of the thermal model.

SHELTER STUDIES

Scaled models of buildings have considerable cost advantages over full size structures for comparison of a number of shelter types at the same experimental site and over the same time period (Hahn *et al*, 1961). There are also cost advantages to using smaller thermal models for measurement of Q_S within model shelters.

Four smaller thermal models of 500 kg cattle were constructed (diameter = 265 mm, length = 570 mm) using a similar design to the larger models (Figure 1). However, the cured hide was replaced by a synthetic material. The synthetic material does not deteriorate and provides a consistent hair coat thermal resistance for the four models. This is difficult to achieve with

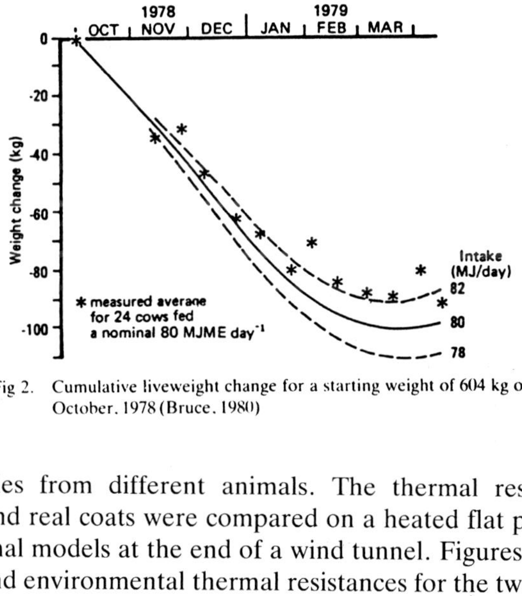

Fig 2. Cumulative liveweight change for a starting weight of 604 kg on 1 October, 1978 (Bruce, 1980)

hide samples from different animals. The thermal resistances of the synthetic and real coats were compared on a heated flat plate and on two small thermal models at the end of a wind tunnel. Figures 3a and 3b show hair coat and environmental thermal resistances for the two materials over a range of windspeeds. The synthetic coat simulates the real coat with good accuracy.

Three simple plywood shelters, 2.5 m square and 1.5 m high, were constructed at the experimental site and consist of

(i) four walls and no roof
(ii) a roof only
(iii) four walls and a roof with ventilation openings simulating a well ventilated cattle building.

A model was placed in each shelter and the fourth was placed in an unsheltered position for comparison with the three shelter treatments and with the large model sited nearby. Heat losses and internal temperatures were automatically logged every hour over 14 weeks between December 1981 and March 1982.

Tables 1a and 1b show ratios of observed Q_s values for each shelter to observed Q_s for the unsheltered small model, calculated hourly and daily respectively.

Tables 2a and 2b show hourly (a) and daily mean (b) ratios of observed Q_s for the small model to observed Q_s for the large model and ratios of predicted to observed Q_s for the large and small unsheltered models respectively.

Table 1 *Ratios of observed Q_s for each shelter to observed Q_s for the unsheltered small model calculated hourly (a) and daily (b).*

	Walls only model	Roof only model	Roof and Walls model
a			
Number of hours	1674	1674	1674
Mean ratio	0.9420	0.9833	0.8006
Standard deviation	0.0354	0.0286	0.0414
Standard error	0.0009	0.0007	0.0010
b			
Number of days (>16 complete hours data)	48	48	48
Mean ratio	0.9358	0.9834	0.8001
Standard deviation	0.0156	0.0156	0.0393
Standard Error	0.0022	0.0023	0.0055

The results show no significant difference between the small and large models and that the predictive model can be applied equally well to both small and large models. The 20% reduction of Q_s by the simulated cattle building is similar to previous results for the large models and to results from Webster *et al* (1970).

Prediction of Q_s using reduction factors for each shelter (from Tables 1a and 1b) may not be applicable to another climate data set, therefore a derivative-free linear regression package (BMDP) was used to derive climate modifiers for each shelter. A comparison of air temperatures inside and outside each shelter showed no significant difference between them.

Table 2 *Hourly (a) and daily mean (b) ratios of observed Q_s for the large model to observed Q_s for the small model, and ratios of predicted to observed Q_s values for both large and small models.*

	Observed large model ÷ observed small model	Predicted ÷ observed (large model)	Predicted ÷ observed (small model)
a			
Number of hours	1524	1524	1524
Mean ratio	0.9994	0.9914	0.9898
Standard deviation	0.0356	0.0465	0.0360
Standard error	0.0009	0.0012	0.0009
b			
Number of days (>16 complete hours data)	48	48	48
Mean ratio	0.9933	0.9982	0.9923
Standard deviation	0.0263	0.0367	0.0214
Standard Error	0.0038	0.0053	0.0031

The remaining climate modifiers were as follows:
For shelter

(i) $0.65 \times R, 1.00 \times r(t), 0.50 \times u$
(ii) $0.28 \times R, 0.80 \times r(t), 1.00 \times u$
(iii) $0.00 \times R, 0.00 \times r(t), 0.10 \times u$.

Tables 3a and 3b give ratios of predicted to observed Q_s values for each shelter and show that using these modifiers in the predictive model results in errors similar to those for the unsheltered models. The coefficient of variation is about 4.0% and 2.5% for hourly and daily predictions respectively.

Figures 4a and 4b show daily mean measured and predicted Q_s values respectively for all models over 24 days.

A further winter's experiment is near completion and it remains to be seen whether these climate modifiers are sufficiently accurate when applied to a different data set. Nevertheless, the results show that given a simple series of climate modifiers the predictive model can be applied to heat losses from sheltered models, and that thermal models have a useful role in deriving and validating these modifiers.

Table 3 *Ratio of predicted to observed Q_s values, calculated hourly (3a) and daily (3b) using climate modifiers derived for each shelter type.*

	Roof only	Walls only	Roof and Walls
a			
Number of hours	1370	1370	1370
Mean ratio	0.9930	1.0009	0.9967
Standard deviation	0.0375	0.0364	0.0340
Standard error	0.0010	0.0010	0.0009
Coefficient of variation	3.8%	3.6%	3.4%
b			
Number of days (>16 complete hours data)	48	48	48
Mean ratio	0.9942	1.0001	1.0011
Standard deviation	0.0226	0.0185	0.0215
Standard Error	0.0031	0.0027	0.0031
Coefficient of variation	2.3%	1.9%	2.2%

Table 4 *Cumulative live-weight deficit for each shelter treatment for the whole winter (28 weeks)*

Shelter	Estimated cumulative live-weight deficit (kg)
No shelter	40
$0.94 \times Q_s$	20
$0.50 \times u$ and $0.65 \times R$	20
$0.80 \times Q_s$	2

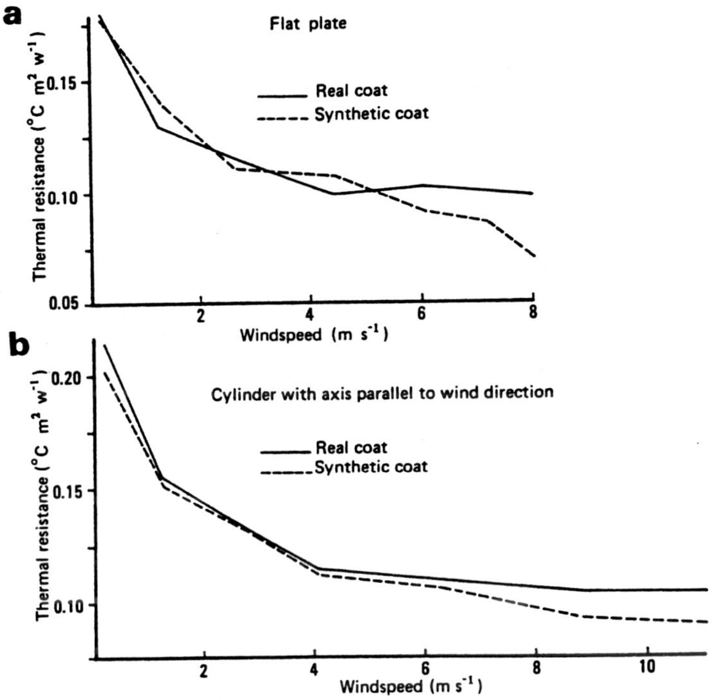

Fig 3. Thermal resistances of coat and air ($I_n + I_a$) plotted against windspeed

APPLICATION

It is generally understood that suckler cows and beef cattle are able to withstand most winter conditions in the United Kingdom with little ill effect when given access to a small degree of shelter (Webster, 1971). It has been shown that, for suckler cows, benefits of housing measured in cow liveweight or feed are greater at low intakes and high liveweight loss but are not likely to exceed a 7 MJ ME day^{-1} saving in feed energy (Bruce, 1982). However, calves are less cold hardy and are either housed or given access to a sheltered creep. The results from the shelter work described can be applied to a predictive equation for the energy balance of such calves. Q_s values for the calves were estimated, using equation 2, from weekly mean values of climate variables measured between October 1978 and April 1979. A number of alterations are required for application of the model to calves. The calf weight increases from 40 kg at birth to 225 kg at 29 weeks by about 0.9 kg day^{-1} with a consequent change in body surface area as defined by

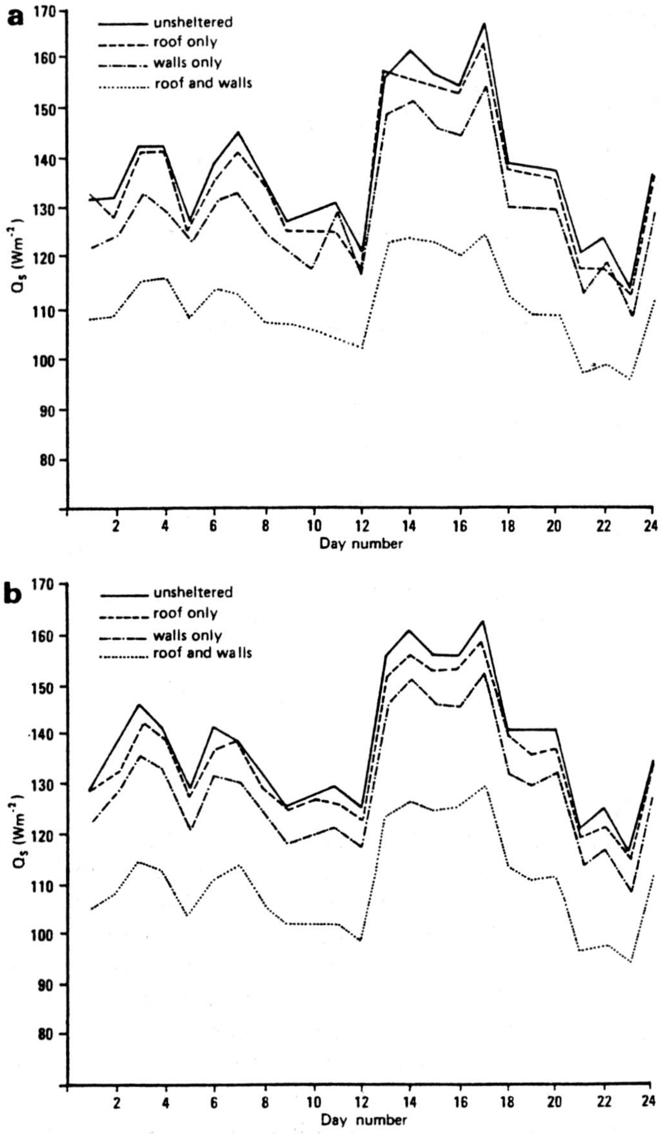

Fig 4. Measured (a) and predicted (b) daily mean Q_s values for each model

$$A = 0.09\,W^{0.67} \tag{6}$$

(Mitchell, 1927).

Vasoconstricted values of I_t were calculated from equation 7 relating I_t to dimensional growth (Bruce & Clark, 1979).

$$I_t = bW^{0.33} \tag{7}$$

The value $b = 0.02$ at birth and rises by increments of 1.37×10^{-4} per week. This results in I_t values of 0.06 °C m^2 W^{-1} at 40 kg (Gonzalez-Jiminez & Blaxter, 1962) and 0.12 °C m^2 W^{-1} at 200 kg (Webster *et al*, 1970). This equation describes the changes of I_t between two measured values at diferent liveweights but is not based on measurements of changes within or outside this liveweight range. Research by Webster *et al* (1976) has shown that I_t increases in relation to body dimension up to 100 kg liveweight. However, further increases are complicated by acclimation to the thermal environment.

Thermoneutral heat production can be estimated from

$$Q_t = 11.57(E_m + \frac{1-k}{k}.E_g)/A \tag{8}$$

where maintainance requirements are a function of liveweight (ARC, 1980). Metabolisable energy is converted to liveweight gain with an efficiency k related to the ratio of total metabolisable energy to total dry matter of the feed (ARC, 1980). Estimates were made of milk, hay, and barley intakes based on the diets fed to cows and calves on experiment (Broadbent, 1981). Fairly large errors in diet estimates give relatively small errors in estimations of k.

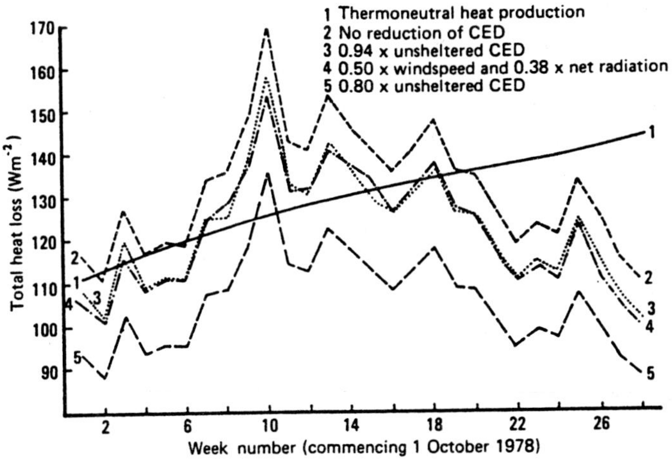

Fig 5. Total heat losses from calves with various degrees of shelter

Estimates of heat losses and thermoneutral heat production are shown in Figure 5.

Heat losses were estimated with no reduction of Q_s for unsheltered calves, with a 20% reduction of Q_s simulating a building, with a 6% reduction of Q_s simulating a walled creep (from Table 1a), and with experimentally derived climate modifiers for a walled enclosure reducing windspeed by 50% and net radiation by 38%. Thermoneutral heat productions were 110 W m^{-2}, 125 W m^{-2} and 140 W m^{-2} for liveweights of 40 kg, 100 kg and 200 kg respectively. These values are similar to measurements by Webster *et al* (1976) who measured heat losses at 5 °C of 123 W m^{-2} and 133 W m^{-2} for liveweights of 100 kg and 200 kg respectively. There is good agreement between heat loss estimates using a 6% reduction factor and those estimated using climate modifiers.

The result of heat losses in excess of thermoneutral heat production can be calculated in terms of reduction of liveweight gain. Table 4 gives estimated cumulative liveweight deficits for each shelter treatment due to redirection of metabolisable energy towards heat production. Measured differences between calves at the same experimental site and calves housed in a nearby building were approximately 22 kg over the same winter (Broadbent, 1981).

These results show that a small reduction of heat loss by a simple walled structure can halve the reduction of liveweight gain due to the climate. This demonstrates the importance of evaluating these relatively small reductions of heat loss in terms of the benefits to the animal.

CONCLUSION

The results from shelter studies using thermal models of cattle have been included in a practical example which indicates the scope for future applications. This technique provides a valuable tool for making shelter decisions for an animal production system in a given geographical area.

REFERENCES

Albright, L.D. (1973). Predicting diurnal building temperature variations by using true models. Journal of Agricultural Engineering Research **18**: 407-410.

A.R.C. (1980). The nutrient requirements of ruminant livestock. Commonwealth Agricultural Bureaux, Slough.

Blaxter, K.L. & Wainman, F.W. (1961). Environmental temperature, energy metabolism, and heat emission of steers. Journal of Agricultural Science **56**: 81-90.

Blaxter, K.L. & Wainman, F.W. (1964). The effect of increased air movement on the heat production and emission of steers. Journal of Agricultural Science **62**: 207-214.

Bond, T.E., Neubauer, L.W. & Givens, R.L. (1976). The influence of slope and orientation on effectiveness of livestock shelters. Transactions of the American Society of Agricultural Engineering, **19**:134-137.

Broadbent, P.J. (1981). Personal communication.

Bruce, J.M. (1978). Natural convection through openings and its application to cattle building ventilation. Journal of Agricultural Engineering Research **23**: 151-167.

Bruce, J.M. & Clark, J.J. (1979). Models of heat production and critical temperature for growing pigs. Animal Production **28**: 353-369.

Bruce, J.M. (1980). Modelling the climatic energy demand on suckler cows. Animal Production **30**: 449-450.

Bruce, J.M. (1982). The effect of feed level and environment on suckler cows. British Society Animal Production Winter Meeting, Paper No. 107.

Burnett, G.A. & Bruce, J.M. (1978). Thermal simulation of a suckler cow. Farm Building Programme **54**: 11-13.

Gonzalez-Jiminez, E. & Blaxter, K.L. (1962). The metabolism and thermal regulation of calves in the first month of life. British Journal of Nutrition **16**: 199-212.

Hahn, L., Bond, T.E. & Kelly, C.F. (1961). Use of models in thermal studies of livestock housing. Transactions of the American Society of Agricultural Engineering, **4**(1): 45-51.

Holmes, C.W. & McLean, N.A. (1975). Effects of air temperature and air movement on the heat produced by young Friesan and Jersey calves, with some measurements of the effects of artificial rain. New Zealand Journal of Agricultural Research **18**: 277-284.

Jordan, W.E., Lister, E.E. & Comeau, J.E. (1969). Outdoor versus indoor winter of fall calving beef cows and their calves. Canadian Journal of Animal Science **49**: 127-129.

Jorgensen, L.J., Jorgensen, N.A., Schingoethe, P. & Owens, M.J. (1970). Indoor versus outdoor calf rearing at three weaning ages. Journal of Dairy Science **53**(6): 813-816.

Koenig, T.J., Hellickson, M.A. & Roepke, W.L. (1978). Building geometry and wind effects on model open-front beef building ventilation. Transactions of the American Society of Agricultural Engineering, **21**: 1199-1208.

McArthur, A.J. & Monteith, J.L. (1980). (1). Boundary layer insulation of a model sheep with and without fleece. Proceedings of the Royal Society of London B **209**: 187-208.

McCarrick, R.B. & Drennan, M.J. (1972). Effects of winter environment on growth of young beef cattle. Animal Production **14**: 93-110.

Mitchell, H.H. (1927). Check formulae for surface areas of animals. Report of the Illinois Agricultural Experimental Station **41**: 155-158.

Perera, M.D.A.E.S. (1981). Shelter behind two-dimensional solid and porous fences. Journal of Wind Engineering Ind. Aerodynamics **8**: 93-104.

Webster, A.J.F., Chlumecky, J. & Young, B.A. (1970). Effects of cold environments on the energy exchanges of young beef cattle. Canadian Journal of Animal Science **50**: 89-100.

Webster, A.J.F. (1971). Prediction of heat losses from cattle exposed to cold outdoor environments. Journal of Applied Physiology **30**(5): 684-690.

Webster, A.J.F., Gordon, J.G. & Smith, J.S. (1976). Energy exchanges of veal calves in relation to body weight, food intake, and air temperature. Animal Production **23**: 35-42.

Wiersma, F. & Nelson, G.L. (1967). Non-evaporative convective heat transfer from the surface of a bovine. Transactions of the American Society of Agricultural Engineering, **10**: 733-737.

APPENDIX

Definition of symbols.

F_i feed intake (MJ ME day^{-1})

E_m energy used for maintainance (MJ ME day^{-1})

E_g energy used for liveweight gain (MJ ME day^{-1})

J energy stored as heat (MJ day^{-1})

Q_e evaporative heat loss (cutaneous and respiratory) (W m^{-2}) (or MJ day^{-1})

Q_s climatic energy demand (CED), or sensible heat loss (W m^{-2}) (or MJ day^{-1})

Q_t thermoneutral heat production (W m^{-2}) (or MJ day^{-1})

T_b internal or deep body temperature of animal (°C)

T_a air temperature (°C)

I_a environmental thermal resistance (°C m^{2}W^{-1})

I_h hair coat thermal resistance (°C m^{2}W^{-1})

I_t tissue thermal resistance (°C m^{2}W^{-1})

R net radiation (W m^{-2})

u windspeed (m s^{-1})

$r(t)$ rain at time t (mm hr^{-1})

N number of hours since last rain
v virtual rain (mm hr^{-1})
k efficiency of conversion of feed metabolisable energy to liveweight gain
A surface area of animal (m^2)
W liveweight of animal (kg)
$\alpha, \beta, \gamma, \delta, a$ = model parameters

Progress in Biometeorology, 1985, Vol. 2, pp. 99-114

WIND AND PLANT PHYSIOLOGY – A REVIEW

D.K.L. MacKerron and P.D. Waister
Scottish Horticultural Institute,
Invergowrie,
Dundee, DD2 5DA.

Shelter alters the microclimate by its effects on wind speed and turbulence, which in turn affect temperature and humidity as well as movement of plant parts. These are the factors that determine the direct physiological response of plants to alteration in the windiness of the environment, but responses to shelter reported in the literature can seldom be attributed only to these factors. Depending upon site, plant species, and nature of the shelter, additional indirect factors may be: pathogens, wind-borne soil, salt or hail, soil erosion, activity of pollinating insects, differential water supply caused by snow drift, shading by the shelter, and competitive effects from the roots of living shelter belts.

It should perhaps be emphasised that some of these indirect factors may be more important in the net effect of shelter on plant growth than are the direct effects we shall review.

During the past 30 years there have been regular reviews of shelter effects. In the early ones (e.g. van Eimern *et al*, 1964) emphasis was on effects on evaporation, but it was subsequently recognised that increase in wind speed may or may not increase transpiration.

The earlier generalisations about effects of wind in lowering tissue temperature have also been modified, and there is more reference to its effects on temperature amplitude. Greater attention has been given to cryptic damage, increase in respiration rates, and to vibration phenomena (thigmomorphogenesis, seismomorphogenesis).

In 1967 Marshall, summarising his review on the effects of shelter on productivity of grasslands and field crops, said 'The increased air temperatures probably operate to increase leaf area, the reduced evaporation increases both net assimilation rate and leaf area'. A survey of wind damage in horticultural crops (Waister, 1972a) concluded that 'further progress in understanding the influence of wind on crops in the field awaits separation of the components of damage', that 'the role played by water relations must still be regarded as uncertain in many situations', and that 'the assessment of growth restriction caused by the effect of wind on tissue temperature is . . . handicapped by a lack of adequate records'.

Under dry conditions on the Great Plains, Rosenberg and his coworkers investigated shelter effects for more than 10 years. By 1975 he could conclude 'Moisture conservation for later plant use is probably the major direct benefit of shelter in dry land agriculture'.

However, in 1977 in a book devoted to plant response to wind, Grace observes that 'As a result of the work reported during the last decade the views of van Eimern *et al* (1964), that the benefit of shelter comes from the greater soil moisture in sheltered places, cannot be accepted'.

It appears that there is still need for critical analysis of the effect of wind on physiological processes.

NET EFFECTS OF WIND ON GROWTH AND DEVELOPMENT

The continuing interest in shelter research in agriculture is understandable in view of the magnitude of reported yield responses, and the fact that wind is one of the weather elements over which the agriculturist can exert some control. Most of the data on crop yield responses to shelter come from field experiments in which growth and yield are compared in sheltered and exposed plots. The results are characterised by variability – between locations, between seasons, between species and even between cultivars of one species. Interpretation of this variability is difficult when so many of the reports include little or no information on the microclimates in the sheltered and exposed areas, or on plant development, other than final yield.

The literature on crop yield responses to shelter from wind has been extensively reviewed (Jensen, 1954; van Eimern *et al*, 1964; Marshall, 1967) and Grace (1977) presents in tabular form a survey of effects on yield reported between 1911 and 1976. Rather than review once more the extensive literature on the subject we refer the reader to these existing reviews.

EFFECTS OF SHELTER ON AIR TEMPERATURE

The first measurements of temperature behind shelterbelts were made in 1872 by La Cour in Denmark (La Cour, 1872; Flensborg & Nøkkentved, 1938). According to him the mean temperature in the protected area is 1.5° C higher during the day than in an exposed area but he recorded lower air temperatures in the protected area at night. Detailed measurements were made by Bodroff (1936) in south Russia and his conclusions, quoted by van der Linde and Woudenberg (1950), were that 'During the first half of the day, when heat-balance is positive, the shelterbelts produce a warming effect. During the second half of the day, when balance is negative, the shelterbelts produce a cooling effect'. These effects can perhaps be best summarised by saying that shelterbelts increase the continental character of the climate by causing greater daily amplitudes of temperature. The more cloud there is the smaller the effect of the shelter on temperature (Bates, 1911).

Van Wijk and Hidding (1955) calculated the amplitudes of air temperature variation at a range of distances from a windbreak. In such work distance is commonly expressed in multiples of *h*, the height of the windbreak. They concluded that, on average, the daily amplitude near the ground could be 1° C greater than in the open for as far as 30 *h* and 2° C greater up to 10 *h* from the windbreak.

These early observations have been confirmed more recently in Nebraska by Rosenberg (1966a) on sugar beet but not on beans (Rosenberg, 1966b). Aase and Siddoway (1974) measured air temperature profiles above winter wheat using shielded thermocouples and found that the influence of windbreaks was not consistent, that during a day profiles in the sheltered area shifted from warmer to cooler and back.

Of course the effects of shelter on air temperature are largely of interest only in default of good measurements of plant tissue temperature and soil temperature.

Aase and Siddoway (1974) found soil temperatures to be initially higher in spring in the sheltered area but enhanced crop growth reduced radiation load on the soil so that temperatures at 5 and 15 cm depths declined in sheltered areas in mid- to late-May while in the exposed areas these temperatures rose. A corollary of this is that more radiation was then being dissipated by the crop itself.

MacKerron (1976a) measured air, soil and tissue temperatures in exposed and sheltered plots of strawberry throughout the year. Air temperatures were very similar in sheltered and exposed plantations but there were differences in both soil and tissue temperatures between the two treatments. Tissue temperatures were measured using small thermistors mounted in fine hypodermic needles. It proved impractical to measure leaf temperatures routinely. Differences were found between exposed and sheltered plants in the temperature of the shoot apex or crown (Figure 1). Throughout the growing season the average temperatures of shoot apices of sheltered plants were higher than those of exposed plants. During the winter the difference disappeared or was reversed. Soil temperatures at 20 cm were consistently higher in the sheltered plots during the summer and consistently lower during the winter (Figure 2). These findings can be explained in terms of the energy balance of the plots. The windbreaks cause reduced advection on the sheltered plots and so in summer the sheltered plots experience slightly higher temperatures. In winter when the radiation balance is negative, the windbreaks are responsible for a reduction in the sensible heat

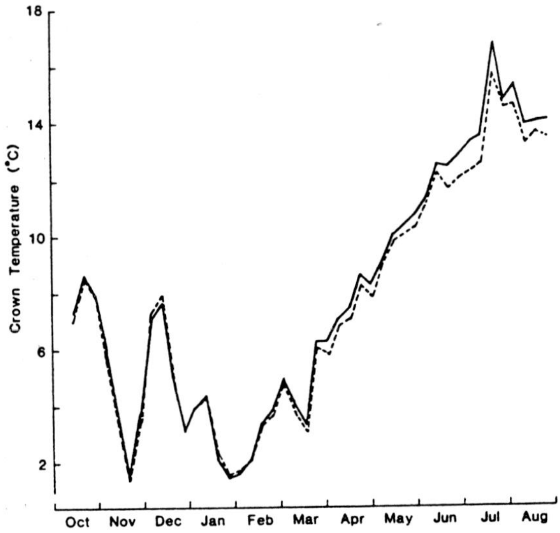

Fig. 1. Weekly mean temperatures of a strawberry crown or shoot apex based on 15-minute measurements on five crowns per treatment. Solid line, sheltered; pecked line, exposed.

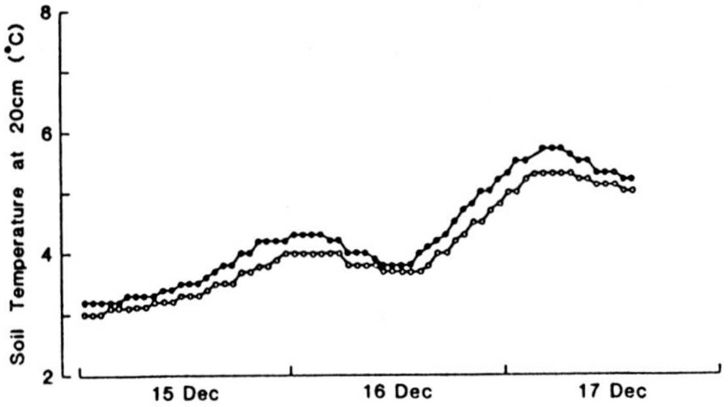

Fig. 2. Soil temperatures at 20 cm in plots of strawberry which were sheltered (open symbols) or exposed (solid symbols).

transferred to the crop which then suffers slightly lower temperature than the exposed crop.

There is a paucity of direct measurement of the effects of wind on tissue temperature *in a crop* but we may perhaps deduce from observations in shelter experiments that, during the growing season, reduced windspeeds will generally lead to higher temperatures and conversely that increased wind speed will lead to lower temperatures or at least will reduce the divergence between tissue and air temperatures. This was predicted by Landsberg *et al* (1974) who gave a theoretical analysis of the energy balance of apple buds in terms of incident radiant energy, air temperature and windspeed such that:

$$\delta T = a\, Q\, u^{-1/2}$$

where a is a constant, Q is the radiation balance of the bud, u is the windspeed and δT is the temperature difference between tissue and the air.

With that particular tissue, loss of latent heat could be assumed to be negligible, and Landsberg *et al* (1974) extended the analysis to apple blossom by neglecting transpiration. The confounded effects of changed boundary layer resistance on sensible heat exchange and latent heat exchange from a surface which is not passive (the stomata may open or close) are the main cause of difficulty in determining the direct effect of wind on tissue temperature.

If we can accept the generalisation that reduced windspeeds during the day may lead to increased temperature then we can make some assessment of the *direct* effects of wind on growth through temperature.

Milthorpe (1959) showed that the rate of leaf appearance in cucumber increased with increasing temperature. Bensink (1971) showed similar results for lettuce and Milford and Riley (1980) showed that in sugar beet the rate of leaf appearance increased linearly with temperature and that the rate was the same in the 9 cultivars examined (rate of emergence = 0.033 leaves $day^{-1}\ °C^{-1}$ above a base temperature of 2° C).

Hay (1978) observed that in east Scotland the mainstem apex of winter wheat is close to the soil surface until late May and that the apex of spring barley did not leave the soil surface until the beginning of June. Until these dates, soil temperature might be expected to modify the rate of development of the crop, since Peacock (1975) found that leaf extension rate in *Festuca* was controlled by stem apex temperature. These results are consistent with the enhanced early growth of winter wheat found by Aase and Siddoway (1974).

A similar response was found by MacKerron & Waister (1972) in strawberry. Plants grown in sheltered plots produced 5 more leaves in the course of each of two growing seasons than did exposed plants so that after two years sheltered plants had 27% more leaves than had exposed plants. Although there was no detectable effect of shelter on final leaf size in strawberry, temperature has been found to affect rates of leaf expansion in a number of species e.g. leaf relative growth rate is linearly dependent on temperature in *Phaseolus vulgaris* (Jones, 1971), in potato (Waister & Ross, 1981) and in a number of other species.

Russel *et al* (1982) followed the development and yield of autumn and spring sown barley and gave data showing green area indices to be linear functions of accumulated temperature up to area indices of about 4.

It may be deduced, therefore, that increased windspeeds and lower temperatures *per se* lead to reduced leaf area formation.

EFFECTS OF WIND ON EVAPORATION AND PLANT WATER RELATIONS

Apart from increases in growth and yield, changes in evaporation rates and alterations of plant water relations are the most frequently reported effects of a windbreak and the associated reduction in wind- speed, although Jensen (1954), Aslyng (1958) and van Eimern *et al* (1964) all indicated that decreased soil moisture use in sheltered areas only occurred during dry years. Skidmore *et al* (1974) found that on days when water stress was low the differences between sheltered and exposed plants of winter wheat were not significant but that on days conducive to plant water stress sheltered plants had lower stomatal resistance, tended to have higher leaf water potential and had photosynthetic rates higher than or equal to those of exposed plants.

The explanations of changes in evaporation in terms of plant responses or of micrometeorology have been varied, have usually been incomplete and not infrequently have involved circular argument. Many studies, particularly the earlier ones, used atmometers or evaporimeters of the Piche type to assess evaporation, although it has been recognised that these instruments do not reflect evaporation from vegetation. This practice led to contradictory statements such as 'Evaporation in the shelter was reduced but . . . (there was) greater transpiration and consequently more rapid withdrawal of soil moisture' (Rosenberg, 1966b).

An increase in wind speed may modify plant water relations through three effects: a decrease in boundary later resistance (r_a), a decrease in the leaf-air temperature difference (δT) and an increase in stomatal resistance (r_s). Whether evaporation from a leaf is increased or decreased by an increase in windspeed depends on the relative changes of δT and of r_a as well as the sympathetic response of the stomata. Brown and Rosenberg (1971) measured exchange coefficients and latent heat flux above sugar beet using gradient measurements and Bowen Ratios. They found that in the morning sheltered crops had higher evaporation rates than exposed crops but that in the afternoon advective heat led to evaporation being higher from the open field. The total daily flux was closely similar in the two areas. Gravimetric determinations of soil water (Brown and Rosenberg, 1972) were in agreement with these findings.

A general statement which can be made is that provision of shelter has the effect of reducing vertical transfer which results in increased water vapour pressure in the air over the sheltered area. Under these conditions transpiration may be reduced and stomata remain more open (lower r_s) (Watts, 1977). Guyot and Seguin (1975) showed that the influence of a windbreak on potential evaporation is small in a maritime climate since the advective contribution to evaporation is less than 20% and since changes in the radiation balance are small. Lomas and Schlesinger (1971) tested the influence of windbreaks on evaporation from Class A pans in areas with and without advective energy contributing to evaporation and compared the results against values predicted by the Penman equation. They found that this equation allowed an accurate estimate of the reduction of evaporation by a windbreak and also indicated whether or not significant reductions should be expected.

Monteith's form (1965) of the Penman equation can be used to predict changes in evaporation rate, when only windspeed is altered.

$$\lambda E = \frac{\Delta H + \rho c_p (e_{s(T)} - e)/r_{aH}}{\Delta + \gamma (r_{av} + r_c)/r_{aH}}$$

Where E is the evaporation rate (g m^{-2}s^{-1}), H is the available energy (W m^{-2}), e is the vapour pressure, $e_{s(T)}$ the saturation vapour pressure at temperature T, r_c is a surface resistance to vapour transfer, r_{av} and r_{aH} are aerodynamic resistances to transfer of vapour and heat respectively and λ, Δ, γ, ρ, c_p are respectively the latent heat of vapourisation of water, the slope of the saturation vapour pressure curve at T, a psychrometric constant, the density of air and the heat capacity of air.

It is not immediately obvious what will be the effects of altered windspeed although the equation can be manipulated with some simplifying assumptions to show that:
if $\beta > \gamma/\Delta$ then evaporation is decreased in increased wind;
if $\beta = \gamma/\Delta$ then evaporation is independent of windspeed;
and if $\beta < \gamma/\Delta$ then evaporation increases in increased wind where β is the Bowen Ratio ($= \lambda E/(H - \lambda E)$). Even this simplification is not unduly

revealing to someone who is not a micrometeorologist, and so we fitted parameters representative of a number of crops into this equation and used actual local weather conditions to calculate evaporation response.

We chose the period 1 April 1982 to 30 September 1982 and for each day used the average solar radiation level, the solar radiation of hour 09, and the highest hourly value. For each radiation level we calculated evaporation rate at the average windspeed of the day and at that speed plus and minus 1 m s^{-1}. The crop parameters which we used are given in table 1, and the results are summarised in figure 3.

Table 1 *Crop parameters*

Crop	d/h	z_o/h	h	r	AI
Potato	0.78	0.04	0.75	2.5	5.0
Bean	0.75	0.08	1.20	2.5	4.0
Cereal	0.75	0.10	0.80	1.7	2.5
Sitka spruce	0.80	0.03	8.00	10.0	9.6

Where d/h and z_o/h are zero plane displacement and roughness lengths respectively as fractions of crop height h is crop height (m), r_s is stomatal diffusion resistance (cm s^{-1}) and AI is green area index.

At the lower radiation levels increased windspeed almost always increased the estimated evaporation and only at the highest radiation levels were there a large number of occasions when increased windspeed led to reduced evaporation.

There would be obvious dangers in extrapolating from this excercise on particular conditions to the general – the weather elements used were those from one site on the east of Scotland in one particular growing season and the values of r_s were kept constant – but nonetheless if one wishes to assess the likelihood of a condition one must test the theory with real weather in real proportions.

In most studies of evaporation from vegetation the assumption is made that conductances of mass, momentum and heat are similar. The possibility of error in that assumption is seldom mentioned and only occasionally is tested. Thom (1971) first put forward the 'shelter-factor' concept *viz* the effect of mutual interference between leaves resulting in lower windspeeds over the leaves and hence less efficient leaf-air exchange. This was expressed in quantitative terms by Landsberg and Thom (1971) and was used for analysis of momentum balance in a coniferous forest by Landsberg

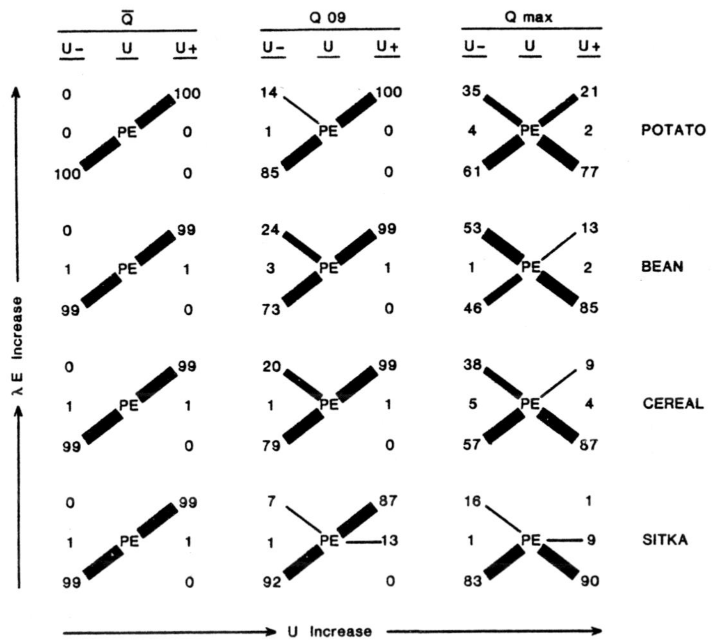

Fig. 3. Schematic representation of frequencies of calculated changes in potential evaporation rates when the windspeed is changed independently of other variables. The numbers indicate the frequency (as a percentage of all occasions) of increased, unchanged or decreased evaporation rates with either change in windspeed. Thus, for the potato crop at $\bar{Q}$ a decrease in windspeed led to decreased calculated evaporation rate on all occasions and an increased windspeed led to an increased calculated rate on all occasions. $\bar{Q}$, daily mean irradiance; Q_{09}, mean irradiance for the hour beginning 0900; Q_{max}, highest hourly mean irradiance; u, mean windspeed; $u-$, a decrease of 1 m s^{-1}; $u+$, an increase of 1 m s^{-1}. The bars indicate the percentage of all occasions: — <20%; ▬ <50%; ■ >50%.

and Jarvis (1973). Landsberg and Powell (1973) developed this to derive shelter factors for drag (p_d) and vapour exchange (p_v). At low windspeeds (about 1 m s^{-1}) the shelter factors for mass and momentum exchange were similarly affected by leaf density but at higher windspeeds p_v was greater than p_d. Transfer resistances for isolated leaves expressed in terms of windspeed and leaf dimension gave results very similar to those obtained by an expression given by Monteith (1965). When leaves were subject to significant mutual interference, resistances to mass transfer were about 50% higher. This emphasises the possibility that exchange coefficients calculated for single leaves, single plants and even small groups of plants in a wind tunnel may not reflect the correct values for a growing crop in the field.

Returning to an assessment of the direct effect of windspeed on plant growth through water status, we may say that in such wind conditions as lead to reduced evaporation, leaf water potentials and turgor will be higher and stomatal conductances are likely to be greater. Where evaporation is increased the reverse will be true.

Frank, Harris & Willis (1974) showed that in irrigated soybean, shelter from wind resulted in a more favourable plant water status while under conditions of limited soil moisture there was enhanced early growth in sheltered plants causing early depletion of soil moisture so that later plant water status was similar in sheltered and exposed plants. Carr (1970) found similar effects in tea in a comparison of the effects of shelter in dry and rainy seasons.

Not only is the influence of shelter on plant water status variable, but the effect on leaf or crop growth is not readily quantifiable. Boyer (1968) and others since have shown that leaf growth is affected by leaf water potential or more strictly turgor potential but for reasons discussed by Hsiao *et al* (1976) simple expressions relating growth and turgor have limited applicability. Yet, one of the most consistent responses of plants to a reduction in wind is an increase in vegetative growth in terms of both leaf area and dry weight (van Eimern *et al*, 1964; Marshall, 1967; Grace, 1977).

This leaves the third major effect of wind on plants *viz* mechanical damage. Early reports of wind damage to plants were concerned with the evident large scale effects of leaf loss and stem breakage so that van Eimern *et al* (1964) did not consider the effects of wind damage on crops and Marshall (1967) mentioned work which identified threshold windspeeds above which damage was caused (Bernbeck, 1954; Tsuboi, 1961). However, the work of Koomen (1957) suggested that no such threshold existed. In his 1972 review, Waister suggested that physical damage at a microscopic level caused either by the mutual abrasion of plant parts or by their repeated flexing in wind was a relatively unexplored field. He later (1972b) implicated this form of damage in the response of strawberries to shelter and there have been a number of subsequent studies of the phenomenon (e.g. Thompson, 1974; Grace, 1974; MacKerron, 1976a,b). Subsequent work by Grace and coworkers has aimed at identifying the changes induced in leaves by exposure to wind, including surface and mechanical effects. Their findings have been interesting but sometimes contradictory. Thus over a

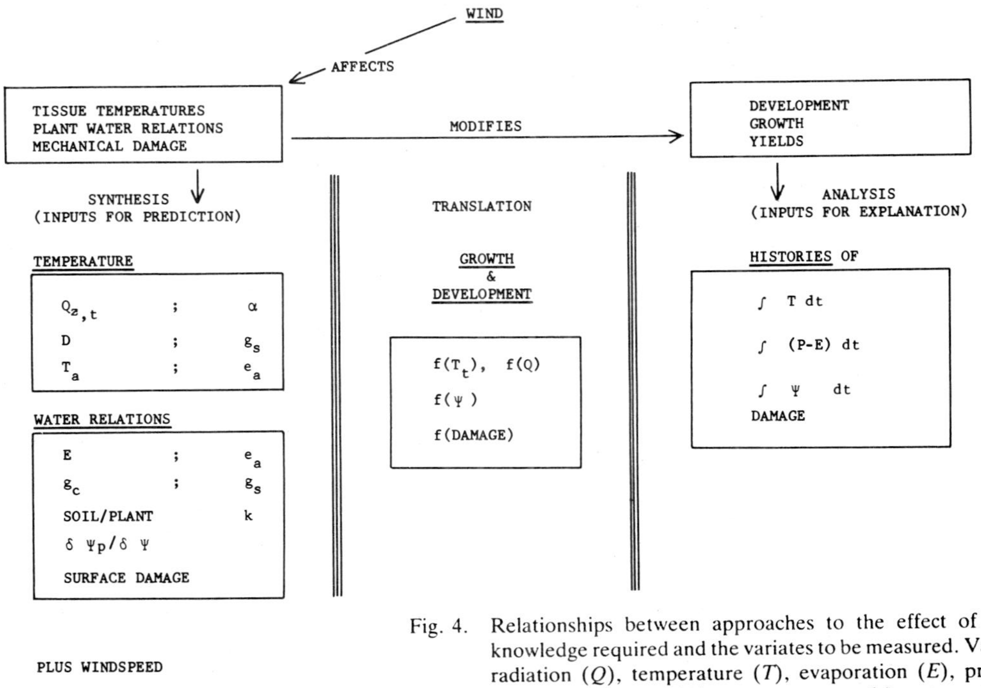

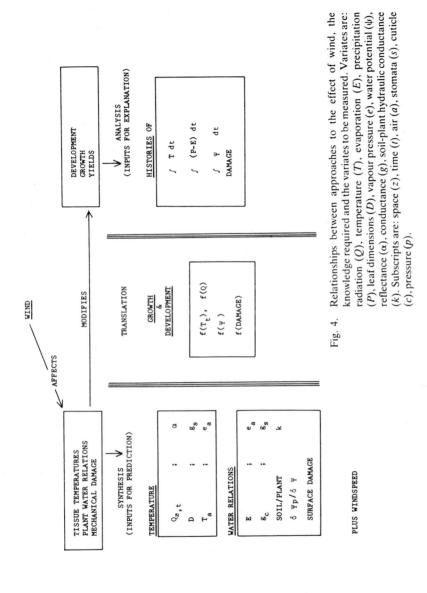

Fig. 4. Relationships between approaches to the effect of wind, the knowledge required and the variates to be measured. Variates are: radiation (Q), temperature (T), evaporation (E), precipitation (P), leaf dimensions (D), vapour pressure (e), water potential (ψ), reflectance (α), conductance (g), soil-plant hydraulic conductance (k). Subscripts are: space (z), time (t), air (a), stomata (s), cuticle (c), pressure (p).

range of species stomatal resistance has been found to increase (Grace *et al*, 1975), decrease (Grace, 1974) or be unchanged (Russell & Grace, 1978a; Rees & Grace, 1981). Photosynthesis has been decreased (Grace & Thompson, 1973) or unchanged (Russell & Grace, 1978a) by exposure to wind. Cuticular resistance has been decreased (Grace, 1974) or unchanged (Russell & Grace, 1978a,b). Solute potential has been decreased (Grace & Russell, 1977) or unchanged (Rees & Grace, 1981), and leaf anatomy and morphology have been adapted (Grace & Russell, 1977) or unchanged (Russell & Grace, 1978a). Two common themes have emerged, however, (a) growth is impaired under windy conditions and (b) mechanical injury or imposed movement are implicated in the growth response.

The phenomenon of imposed movement modifying development has been termed thigmomorphogenesis by Jaffe (1973) who showed that rubbing reduced stem elongation in *Phaseolus vulgaris* and that flexure of the stem by wind alone caused reductions which increased with the angle of flexion (Jaffe, 1976) and produced plants more able to withstand high winds. However, Jaffe and Biro (1979) emphasised that in nature all plants must undergo thigmomorphogenesis because wind is a ubiquitous component of the environment. Therefore glasshouse grown plants are the abnormal form.

What is required is a quantitative treatment of thigmomorphogenesis of plants which are already adapted to or modified by slight but continual perturbation. The state of our knowledge of the effects of wind on the physiology of plants can be best summarised, perhaps, by reference to Figure 4.

We know that wind affects tissue temperatures and plant water relations and induces mechanical damage and that somehow these effects are integrated into modified development, growth and yield.

Research programmes on the effects on yield are generally concerned either with synthesis *or* with analysis.

Those workers seeking to predict the effect of wind on temperature or water relations need to know many variables and their interactions. To synthesise the effects of these factors into a prediction of growth requires a greater knowledge of the integrated effects of tissue temperature, radiation, water status and damage than is currently available.

On the other hand those concerned with the net effects of wind on growth may record the histories of environmental variables but lack the framework with which to integrate the effects of the variables so that their data tend to be selective and inadequate for quantitative analysis of response.

Closing the gap between the predictive and explanatory approaches with confidence and reliability will require a recognition of the limitations of each approach and a joint effort at understanding.

REFERENCES

Aase, J.K. & Siddoway, F.H. (1974). Tall wheatgrass barriers and winter wheat response. Agricultural Meteorology **13**: 321-338.

Aslyng, H.C. (1958). Shelter and its effect on climate and water balance. Oikos **9**: 282-310.

Bates, C.G. (1911). Windbreaks: their influence and value. Bulletin No. 86 US Forestry Service pp 100.

Bensink, J. (1971). On the morphogenesis of lettuce leaves in relation to light and temperature. Mededelingen van de Landbouwhogeschool Wageningen **71**: 15.

Bernbeck, O. (1954). Wind und physiologische Teifgründigkeit in ihrer Bedeutung für die Bodenkultur. Berlin: Deutsch. Bauerverlag, pp 100.

Bodroff, V. (1936). The influence of shelter belts on the microclimate of adjacent territories. Journal of Forestry **34**: 696-697.

Boyer, J.S. (1968). Relationship of water potential to growth of leaves. Plant Physiology **43**: 1056-1062.

Brown, K.W. & Rosenberg, N.J. (1971). Turbulent transport and energy balance as affected by a windbreak in an irrigated sugar beet (*Beta vulgaris*) field. Agronomy Journal **63**: 351-355.

Brown, K.W. & Rosenberg, N.J. (1972). Shelter-effects on microclimate, growth and water use by irrigated sugar beets in the Great Plains. Agricultural Meteorology **9**: 241-263.

Carr, M.K.V. (1970). The role of water in the tea crop. In Physiology of Tree Crops. L.C. Luckwill & C.V. Cutting (Eds) pp 287-305. London: Academic Press.

van Eimern, J., Karschon, R., Razumova, L.A. & Robertson, G.W. (1964). Windbreaks and Shelterbelts. Technical Notes, World Meteorological Organisation, Geneva No. 59, pp 188.

Flensborg, C.E. & Nøkkentved, C.S. (1938). Laevirkningsundersøgelser og typebestemmelser of laehegn. Hedeselskabets Tidsskrift **59**: 75-142.

Frank, A.B., Harris, D.G. & Willis, W.O. (1974). Windbreak influence on water relations, growth and yield of soybeans. Crop Science **14**: 761-765.

Grace, J. (1974). The effect of wind on grasses I. Cuticular and stomatal transpiration. Journal of Experimental Botany **25**: 542-551.

Grace, J. (1977). Plant Responses to Wind. London: Academic Press.

Grace, J., Malcolm, D.C. & Bradbury, I.K. (1975). The effect of wind and humidity on leaf diffusive resistance in Sitka spruce seedlings. Journal of Applied Ecology **12**: 931-940.

Grace, J. & Russell, G. (1977). The effect of wind on grasses III. Influence of continuous drought or wind on anatomy and water relations in *Festuca arundinacea* Schreb. Journal of Experimental Botany **28**: 268-278.

Grace, J. & Thompson, J.R. (1973). The after-effect of wind on the photosynthesis and transpiration of *Festuca arundinacea*. Physiologia Plantarum **28**: 541-547.

Guyot, G. & Seguin, B. (1975). Modification of land roughness and resulting microclimatic effects: a field study in Brittany. In Heat and Mass Transfer in the Biosphere Vol. 1. de Vries & Afgan (Eds) pp 467-478 Washington: Scripta Book Co.

Hay, R.K.M. (1978). Seasonal changes in the position of the shoot apex of winter wheat and spring barley, in relation to the soil surface. Journal of Agricultural Science, Cambridge **91**: 245-248.

Hsiao, T.C., Acevedo, E. Fereres, E. & Henderson, D.W. (1976). Water stress, growth, and osmotic adjustment. In A discussion on water relations of plants. J.L. Monteith & P.E. Weatherley (Eds) Philosophical Transactions of the Royal Society of London B **273**: 479-500.

Jaffe, M.J. (1973). Thigmomorphogenesis: the response of plant growth and development to mechanical stimulation. Planta (Berlin) **114**: 143-157.

Jaffe, M.J. (1976). Thigmorphogenesis: a detailed characterisation of the response of beans (*Phaseolus vulgaris* L.) to mechanical stimulation. Zeitschrift für Pflanzenphysiologie **77**: 437-453.

Jaffe, M.J. & Biro, R. (1979). Thigmorphogenesis: the effect of mechanical perturbation on the growth of plants, with special reference to anatomical changes, the role of ethylene and inter- action with other environmental stresses. In Stress Physiology in Crop Plants, H. Mussell & R. C.Staples (Eds) pp 25-59. New York: Wiley - Interscience.

Jensen, M. (1954). Shelter effect: investigations into the dynamics of shelter and its effect of climate and crops. Copenhagen: Danish Technical Press. pp 264.

Jones, L.H. (1971). Adaptive responses to temperature in dwarf french beans, *Phaseolus vulgaris* L. Annals of Botany **35**: 581-596.

Koomen, J.P. (1957). Experiences with shelter provided by various crops. Mededelingen 6. Proefstr. Groenteteelt volle Grond. pp 43.

La Cour, P. (1872). Skoveness inflydelse paa varmen. Reference in Zeitschrift Österreich.Ges. für Meteorologie **7**: 254-256.

Landsberg, J.J., Butler, D.R. & Thorpe, M.R. (1974). Apple bud and blossom temperatures. Journal of Horticultural Science **49**: 227-239.

Landsberg, J.J. & Jarvis, P.G. (1973). A numerical investigation of the momentum balance of a spruce forest. Journal of Applied Ecology **10**: 645-655.

Landsberg, J.J. & Powell, D.B.B. (1973). Surface exchange characteristics of leaves subject to mutual interference. Agricultural Meteorology **12**: 169-184.

Landsberg, J.J. & Thom, A.S. (1971). Aerodynamic properties of a plant of complex structure. Quarterly Journal of the Royal Meteorological Society **97**: 565-570.

van der Linde, R.J. & Woudenberg, J.P.M. (1950). On the microclimatic properties of sheltered areas: the oak-coppice sheltered area. Kon. Nederl. Met. Inst. Mededel. en Verhandel. No. 102, Ser. A **56** pp 151.

Lomas, J. & Schlesinger, E. (1971). The influence of a windbreak on evaporation. Agricultural Meteorology **8**: 107-115.

MacKerron, D.K.L. (1976a). Shelter for strawberries. Proceedings 4th Symposium on Shelter Research (MAFF, Warwick) pp 13-22.

MacKerron, D.K.L. (1976b). Wind damage to the surface of strawberry leaves. Annals of Botany **40**: 351-354.

MacKerron, D.K.L. & Waister, P.D. (1972). Crop response to shelter. Annual Report Scottish Horticultural Research Institute for 1971.

Marshall, J.K. (1967). The effect of shelter on the productivity of grasslands and field crops. Field Crop Abstracts **20**: 1-14.

Milford & Riley (1980). The effects of temperature on leaf growth of sugar beet varieties. Annals of Applied Biology **94**: 431-443.

Milthorpe, F.L. (1959). Studies on the expansion of leaf surface. I The influence of temperature. Journal of Experimental Botany **10**: 233-249.

Monteith, J.L. (1965). Evaporation and environment. Symposium of the Society for Experimental Biology 19: 205-234.

Peacock, J.M. (1975). Temperature and leaf growth in *Lolium perenne*. II. The site of temperature perception. Journal of Applied Ecology **12**: 115-123.

Rees, D.J. & Grace, J. (1981). The effect of wind and shaking on the water relations of *Pinus contorta*. Physiologia Plantarum **51**: 222-228.

Rosenberg, N.J. (1966a). Influence of snow fence and corn windbreaks on microclimate and growth of irrigated sugar beets. Agronomy Journal **58**: 469-475.

Rosenberg, N.J. (1966b). Microclimate, air mixing and physiological regulation of transpiration as influenced by wind shelter in an irrigated bean field. Agricultural Meteorology **3**: 197-224.

Rosenberg, N.J. (1975). Windbreak and Shelter Effects. In Progress in Biometeorology pp 108-134. Lisse: Swets and Zeitlinger.

Russell, G., Ellis, R.P., Brown, J., Milbourn, G.M. & Hayter, A.M. (1982). The development and yield of autumn- and spring-sown barley in south east Scotland. Annals of Applied Biology **100**: 167-178.

Russell, G. & Grace, J. (1978a). The effect of wind on grasses IV. Some influences of drought or wind on *Lolium perenne*. Journal of Experimental Botany **29**: 245-255.

Russell, G. & Grace, J. (1978b). The effect of wind on grasses V. Leaf extension, diffusive conductance, and photosynthesis in the wind tunnel. Journal of Experimental Botany **29**: 1249-1258.

Skidmore, E.L., Hagen, L.J., Naylor, D.G. & Teare, I.D. (1974). Winter wheat response to barrier-induced microclimate. Agronomy Journal **66**: 501-505.

Thom, A.S. (1971). Momentum absorption by vegetation. Quarterly Journal of the Royal Meteorological Society **97**: 414-428.

Thompson, J.R. (1974). The effect of wind on grasses II. Mechanical damage in *Festuca arundinacea* Schreb. Journal of Experimental Botany **25**: 965-972.

Tsuboi, Y. (1961). Ecological studies on rice plants with regard to damages caused by wind. Bulletin National Institute for Agricultural Science, Tokyo Ser. A **8**: 1-56.

Waister, P.D. (1972a). Wind damage in horticultural crops. Horticultural Abstracts **42**: 609-615.

Waister, P.D. (1972b). Wind as a limitation on the growth and yield of strawberries. Journal of Horticultural Science **45**: 435-445.

Waister, P.D. & Ross, H.A. (1981). Seasonal changes in net assimilation rate and relative growth rate of stem cuttings of potato. Potato Research **24**: 221.

Watts, W.R. (1977). Field studies of stomatal conductance. In Environmental Effects on Crop Physiology J.J. Landsberg & C.V. Cutting (Eds) pp 173-189. London: Academic Press.

Wijk, W.R. & Hidding, A.P. (1955). Onderzoek over de verandering van het klimaat achter windschermen. Landbouwkdg. Tijdschrift, (s' Gravenhage) **67**, No. 10: 707-712.

Progress in Biometeorology, 1985, Vol. 2, pp. 115-126

WIND AND SURFACE DAMAGE

C.E.R. Pitcairn and J. Grace
Department of Forestry and Natural Resources,
University of Edinburgh,
King's Buildings,
Edinburgh, EH9 3JU.

INTRODUCTION

It is quite clear by now that there are at least three main ways in which wind may influence plant growth. The first and perhaps most intriguing, came to light after the publication of a report on the effect of shaking on the growth of a woody plant, *Liquidambar styraciflua* (Neel & Harris, 1971). This was one of the first demonstrations that plant growth, in particular extension growth, is inhibited by mechanical stimuli. Even before then, it was well known that the distribution of growth in trees is markedly influenced by motion, as staked trees develop height growth at the expense of radial growth (Larson, 1965). The underlying mechanism of such reactions is at present unknown, but under investigation by Jaffe and his co-workers (e.g. Erner & Jaffe, 1983). Whatever the mechanism, it is widely held that wind reduces height growth in plants through the motion it causes.

The second way in which wind affects the performance of plants is through an influence on the energy balance of the leaves and meristems. Since this is a purely physical phenomenon, brought about by convective transport of heat and matter, it ought to be possible to calculate the magnitude of the effect on surface temperatures and rates of water loss. This subject was addressed by at least two authors in the 1960's, who discussed the relationship between wind and heat and water vapour transport (Linacre, 1964; Monteith, 1966). Progress in this field depends on a knowledge of the radiative and aerodynamic properties of leaves and canopies. Several authors have attempted to calculate the effect of wind on leaf temperature and transpiration rate (Gates & Papian, 1971; Taylor, 1975; Campbell, 1977; Grace, 1983), with conclusions which vary somewhat according to the original assumptions being made. There is however a general consensus that wind causes a decline in the transpiration rate of sunlit leaves along with a lowering of the leaf temperature. Experiments conducted with well-illuminated plants support this view (Yamaoka, 1958; Drake *et al*, 1970; Dixon & Grace, 1984).

Nevertheless, it is a matter of common observation in horticulture that wind causes 'scorching' or 'burning' of leaves, apparently as a result of dessication, and especially in broad leaved species like *Phaseolus vulgaris* and *Solanum tuberosum*. This brings us to the third main way in which wind influences plant growth, and the subject of this review.

It seems almost certain that such phenomena are the result of surface damage to the leaf; either through repeated flexing of the leaf or by mutual abrasion of leaves in bunches. Such damage is liable to substantially reduce diffusive resistance and lead to enhanced rates of water loss.

Before reviewing this subject, it will be necessary to recall the structure of the leaf surface to identify the nature and whereabouts of the barriers to water loss.

SURFACE DAMAGE

Leaf surface structure

The epidermes of aerial organs of higher plants and ferns are usually covered by a continuous cuticular layer, upon which epicuticular waxes may be superimposed. The cuticle is tightly attached to the tangential walls of the epidermal cells and may extend in a thinner form into part of the substomatal cavities.

Cuticles are effective water barriers and many workers have attempted to isolate and describe the nature of the barrier (Fig. 1). Roelofsen (1952)

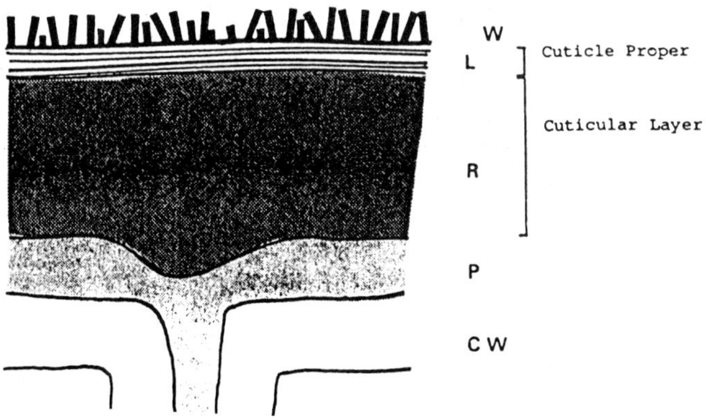

Fig. 1. The structure of the plant cuticle.P, pectin layer and middle lamella; CW, cell wall consisting of alternate layers of cellulose fibres and layers containing, predominently, hemicellulose and pectin; R, reticulate region of cuticle, cutin and wax traversed by cellulose fibrils also called cuticular layer; L, lamellate region of the cuticle, separate lamellae of cutin and wax also called cuticle proper; W, the epicuticular wax. (After Juniper & Jeffree, 1983).

described the outer epidermal wall as consisting of a cuticular membrane and a cell wall, and found that in many species, the membrane was composed of a thin, outer cuticle proper and a thicker cuticular layer as in *Clivia* and *Allium* bulb scales. Such a distinction could not be made for other species studied.

Cuticles are heterogeneous membranes, the components of which were classified by Schonherr (1982) according to their solubility in lipid solvents such as chloroform. The bulk of the membrane is represented by the insoluble fraction or polymer matrix which has a high water permeability due to the presence of strong dipoles which are hydrated in the presence of water. The permeability of cuticles to water vapour is found to be determined by the chloroform-soluble fraction, called the soluble cuticular lipids (waxes) because of the presence of long hydrophobic chains which are unbranched and highly orientated. This complex lipid mixture constitutes only a small fraction (2-30% by weight) of the total membrane mass and hence water permeability is not proportional to membrane thickness or weight. Structurally, it is likely that the cuticle proper, although thin, is the main barrier to permeability because of the presence of soluble cuticular lipids, and there is evidence of a lamellate structure. Juniper & Jeffree (1983) described the 'cuticle proper' of Roelofsen (1952) as the lamellate region of the cuticle composed of separate lamellae of wax (soluble cuticular lipids) and cutin. The lamellar structure of *Citrus* cuticle was shown to be composed of distinct layers of hydrophobic soluble cuticular lipids in the solid state, orientated parallel to the membrane surface. Wattendorf and Holloway (1980) found that *Agave* cuticle proper was composed of layers of electron-opaque and electron-translucent layers. Cutin was the major component of the opaque layer (and most of the underlying cuticular layer) and the translucent lamellae were tentatively ascribed to wax. However *Agave* leaves contain much more wax than could be accommodated in the electron-translucent layers of the cuticle proper. Similar lamellae were described in the cuticle proper of *Apium, Eryngium* and *Humulus* but were not seen in *Abutilon* or *Rumex* (Chafe and Wardrop, 1973).

The relatively permeable cuticular layer, (reticulate region in Fig. 1) chiefly composed of polymer matrix, may provide a mechanically resistant matrix for soluble liquids to be embedded in (Schonherr, 1982) and may also act as a cushion, absorbing strains caused by flexure (Juniper & Jeffree, 1983). Evidence suggests that the electron-dense fibrillae of the cuticular layer are anastomosing channels of high polarity connecting with the cuticle proper and serving the function of a pathway for cuticular precursors and epicuticular waxes (Martin & Juniper, 1970; Merida *et al*, 1981). The soluble cuticular lipids were assumed to be located within electron-lucent globules of the cuticular layer in a highly ordered state but their exact location and orientation within the cuticle is not yet known.

The majority of plant species have wax embedded in or exuded over the cuticle surface, the so-called epicuticular wax. Some plants have a prominent waxy bloom due to the reflection and scattering of light on the surface by wax crystals of diverse form whose dimensions are close to or a little above the wave-length of light (Juniper & Jeffree, 1983; Martin & Juniper, 1970). Some cuticles may have no bloom but still be very rich in wax as in olive and lemon leaves. The role of epicuticular waxes alone in permeability is uncertain as it is difficult to remove them without disturbing the intracuticular soluble cuticular lipids.

Microscopic damage after exposure to wind

All plant surfaces are subject to weathering. Wind causes adjacent leaves to rub against each other creating various kinds of damage. Thomson (1974) grew *Festuca arundinacea* in a wind tunnel at high wind speed (3.5 m s^{-1}). More damage occurred on the strongly ridged adaxial surface than the less structured abaxial surface. Surface cells on ridges were torn open and there appeared to be a greater accumulation of waxes between ridges.

Strawberry leaves developed large brown lesions as a result of wind damage (MacKerron, 1976) and SEM examination of these showed smoothing and disruption of surface waxes and collapse of periclinal walls of the epidermis. Wilson (1978, & in press) investigated the microscopic effects of wind damage on leaves of *Acer pseudoplatanus* with SEM and light microscope. As in strawberry, the abaxial surface sustained the greatest damage developing light- and dark-brown lesions. Examination of the latter showed disruption of fine, epicuticular waxes; flattening of cells and papillae and collapse of epidermal cells and spongy mesophyll were also observed. Samples of *Molinia caerulea* adaxial epidermis were examined with SEM after low and high wind speed treatment in a wind tunnel (Pitcairn & Grace, in press). The low windspeed treated material showed considerable wax deposits (Fig. 2) but at high wind speed, the wax had been smoothed and scoured and outlines of silica containing cells were visible. Actual rupture of epidermal cells as observed in *F. arundinacea* by Thomson (1974) was not seen, but some cracking around the edges of the silica bodies was apparent.

Much research has centred around the damage caused by wind blown sand or soil particles. Such damage is far more extreme than that already described, plant tissue blackening rapidly and cells being rapidly crushed on the windward side of the tissue (Fryrear, 1971; Armbrust, 1974).

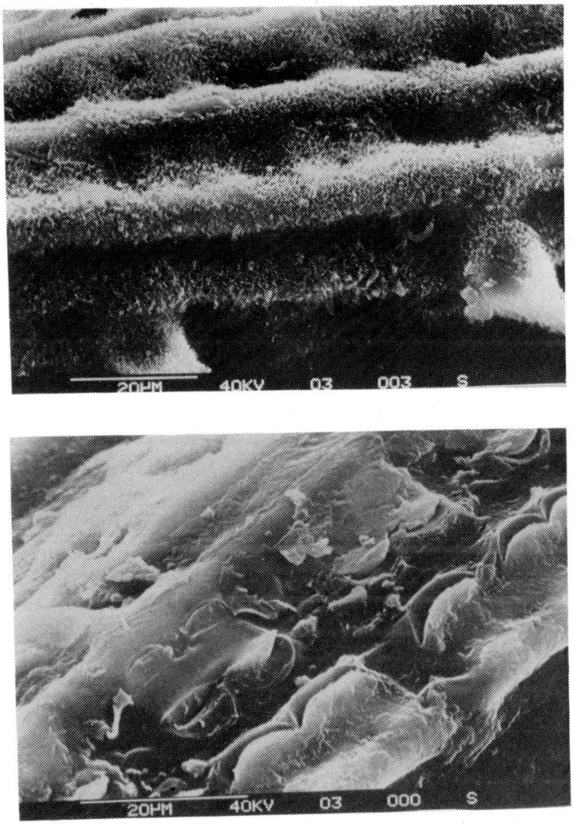

Fig. 2. Scanning electron micrographs of the adaxial leaf surface of *Molinia caerulea*. The upper micrograph is of leaf material grown in the absence of wind, the lower shows material grown at high wind speed ($7.4\,m\,s^{-1}$).

EFFECTS OF DAMAGE ON SURFACE CONDUCTANCE

Transpiration rate is usually controlled by the diffusive conductance of the stomatal pores (Weatherley, 1965) in parallel with that of the cuticle which normally remains unchanged. Cuticular conductance is usually much smaller than that of the stomata and can generally be ignored except when a limited water supply is available to the plant or when cuticle damage results in water loss.

Grace (1974) estimated stomatal and cuticular transpiration rates in grasses by the method first used by Hygen (1951). Measurements were made before and after a 36 hour period of high wind speed (3.5 m s^{-1}) in a wind tunnel. Wind was found to reduce both cuticular and stomatal resistance. The former was attributed to increased numbers of collisions between neighbouring leaves resulting in surface abrasion and tearing similar to that observed by Thomson (1974). The reduction in stomatal resistance was more difficult to explain as the concomitant reduction in leaf water content might be expected to lead to stomatal closure and an increased resistance to water loss. However, Heath (1939) found that passive stomatal opening could occur when subsidiary cells were punctured and it is possible that loss of turgor pressure caused by reduction in water supply or indeed wind damage, could cause guard cells to bulge into flaccid neighbouring cells and spring open (Stalfelt, 1955; Raschke, 1970).

The effect of wind on water loss may depend on leaf age as found by Wilson (1968) for *Acer pseudoplatanus* grown in a wind tunnel. The high day-time leaf conductance of young leaves was little affected by increased wind speed, whereas the low day-time leaf conductance of fully expanded leaves and low night conductances were increased by wind. This appeared to be related to the posture of the leaf, not the cuticular properties. Sycamore leaf damage, measured macroscopically (Wilson, 1978) was correlated with the increase in cuticular conductance although it represented a small proportion of total leaf conductance.

A large degree of the leaf surface damage caused by wind consists of smoothing and scouring of epicuticular waxes, and such damage alone without accompanying rupture or tearing of the surface can also lead to increased water loss. Field observations by Hall & Jones (1961) on *Trifolium repens* showed that contact between leaves and the ground, caused by wind, resulted in much wax removal over the crowns of the epidermal cells at a rate too fast for renewal. They were able to simulate weathering by brushing the leaf surface with a camel hair brush and found an increased water loss in the cuticular phase after stomatal closure. Polishing apples with a soft cloth can result in wax removal and an increase in transpiration (Pieniazec, 1944; Hall, 1966). Others have shown that it is not only the removal of wax that causes increased permeability but also the alteration in the crystalline structure of epicuticular waxes. Commercial dipping of grapes in chemicals to aid drying does not remove significant amounts of wax but results in the formation of a continuous aqueous emulsion over and between wax plates thus greatly aiding diffusion (Chambers & Possingham, 1963). Rather than risk damaging intracuticular waxes by rubbing or brushing, Denna (1970) compared the transpiration rates of glaucous and non-glacous sibling lines of *Brassica oleracea*. In all cases, the non-glaucous plants lost more water through cuticular transpiration than their glaucous siblings, stomatal transpiration rates not being significantly affected. However, when leaf surfaces were rubbed with a cheese cloth both cuticular and stomatal transpiration were increased suggesting that both epicuticular and cuticular waxes are of importance in limiting water loss.

When cuticular waxes were extracted with lipid solvents from astomatous cuticular membranes isolated from *Citrus aurantium* leaves, *Pyrus communis* leaves and *Allium cepa* bulb scales, water permeability was increased by a factor of 300-500 (Schonherr, 1976). Permeability coefficients were completely determined by the waxes (soluble cuticular lipids), the contribution of the polymer matrix being negligible. Hence permeability is independant of cuticle thickness.

Permeability coefficients for cuticular or periderm membranes are variable but very low (Schonherr, 1982), and it would seem that this is the permeability barrier necessary for survival of plants in the terrestrial environment. Thus the difference between species lies not in permeability coefficients but in the manner of achieving them, some plants having thin membranes (*Allium, Pyrus*) others having thick ones (*Clivia, Lycopersicon*). Clearly, soluble cuticular lipids represent the major water barrier, but the amounts of and composition of these lipids differ amongst species of very similar permeabilities.

CAN SURFACE DAMAGE BE REPAIRED?

When plants are wounded, internal tissues become exposed and begin a new function as the plant surface. The characteristic behaviour of tissues after wounding is known as the 'wound response'. Mitotic activity is stimulated and a wound periderm is produced close to the damaged surface. Suberin and lignin are deposited in the outer cell walls of the wound periderm. Cuticle may repair if badly damaged. Lipetz (1970) reported that synthesis of cutin at a wound site involved the activity of three fatty acid oxidising enzymes, stearic acid oxidase, oleic acid oxidase and lipoxidase and one supposes that cuticular abrasion would be followed by renewal of cutin and waxes in a similar way.

Banana leaf tears from which considerable cuticular transpiration occurred (Brun, 1961), eventually became protected from excessive water loss by suberization of tissue along the wound; and a similar 'drying out' of dark brown to light brown lesions in wind-damaged *Acer pseudoplatanus* was observed by Wilson (1978) probably due again to suberization of exposed tissues. Many bloomed plants have the ability to recover from mechanical damage, some such as *Chrysanthemum segetum* even in old age (Martin & Juniper, 1970). When wax was rubbed off *Eucalyptus*, recovery was apparent under EM after three hours and complete after 48 hours (Hallam, 1967). Wax regeneration generally does not restore the wax layer to its pre-damage thickness but represents the remaining wax still to be synthesised rather than a new, extra, synthesis (Juniper & Jeffree, 1983).

How long after abrasion do adverse effects last? *Molinia caerulea* was grown in a wind tunnel at low wind speed (0.7 m s^{-1}) for five days and thereafter at high wind speed (3.5 m s^{-1}) and leaf conductance was measured every 24 hours in dark and light phase. Both stomatal and cuticular conductance increased dramatically (Fig. 3) the day after wind speed had

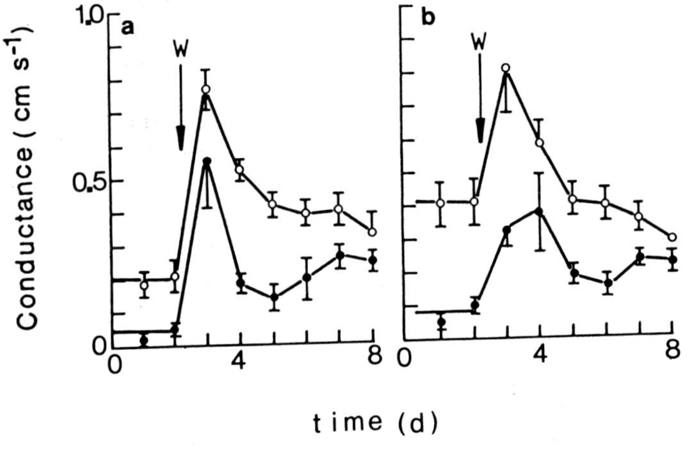

Fig. 3. Changes in surface conductance by day (o) and at night (•) in *Molinia caerulea* following an increase in wind speed from 0.7 m s^{-1} to 8.5 m s^{-1}. The increase in wind speed is indicated by the arrow (W). The results a) and b) are for 2 provenances both collected in West Wales.

been increased but then fell off again and decreased slowly. After five days at high wind speed, stomatal conductance had almost returned to its low wind speed level but cuticular resistance did not return to the low value recorded at low wind speed. If night time conductances of days 5 and 11 are compared, one can see that surface repair occurred but only to a 50% level.

Recent experiments have been carried out in our laboratory to investigate recovery and repair following wind damage. The adaxial surfaces of *Festuca arundinacea* Schreb. were either mechanically abraded using one stroke of fine emery paper or polished using soft tissue. Leaf surface conductances measured every 24 hours were increased by both treatments at night and further increased by a smaller amount in the light. Polishing caused the greater increase in conductance possibly due to both wax removal and alteration of the crystalline structure of the epicuticular waxes as suggested by Chambers & Possingham (1963). Fine mechanical abrasion would have removed some wax and disrupted some cells on the epidermal ridges but tissue and waxes between ridges could have been protected. Future SEM studies should help to elucidate this point. After five days surfaces were repaired to some degree and conductances approached those of the controls, finally levelling off at an elevated value suggesting some degree of permanent damage.

In conclusion, it is evident that the resistance to water vapour diffusion lies in the wax components of the cuticle. One of these components, the epicuticular wax, is delicate and can probably be redistributed even by the shearing force of the wind. The other component in the cuticle proper is much more difficult to remove but seems susceptible to abrasion from other leaves or blown material. The degree of abrasion is likely to be a function of wind speed and the design of the cuticle, but depends also on the posture of the leaves and their mass. The extent of repair, and its time course, has not been studied very much yet; but we do at least know that repair is possible.

Clearly, there is likely to be much specific variation in all these attributes.

CONSEQUENCES FOR PLANT WATER STATUS

Any increase in surface conductance causes an increase in transpiration rate. This increase in transpiration may not however be in strict proportion to the increase in surface conductance, as an increase necessarily affects the energy balance of the leaf, cooling the surface, thus causing a reduction in the saturated vapour pressure within the substomatal cavities.

The relationship between transpiration and water stress has been reviewed recently by Jones (1983). In nearly all species studied an increase in transpiration causes a reduction in leaf water potential. Often this relationship is linear, consistent with the view of the plant as a fixed hydraulic resistance. Sometimes, the relationship is markedly curvilinear, implying a change in resistance as flow through the plant increases. In cases involving bulky tissues, as in trees, there is a large degree of hysteresis in this relationship, suggesting the withdrawal of water from a store.

It is possible that this relationship itself depends on the wind regime to which the plant has been exposed. There are many reports in the literature suggesting a redistribution of tissue types in plants raised in the wind (Grace & Russell, 1982). Such a redistribution may subtly shift the proportion of xylem to epidermis, and thus greatly affect the relationship. Recently, we investigated this relationship in contrasting provenances of *Molinia caerulea* which were known to be susceptible or resistant to exposure to high wind. In this case the susceptible provenance differed from the resistant not in the attributes of the cuticle but in the relationship between transpiration and water potential (Pitcairn & Grace, in press).

Nevertheless, in the more general case it is likely that the structure of the cuticle determines the probability of abrasive damage and the incidence of so-called 'scorching' or 'burning' following gales. It is also likely that at least part of the benefit to crops of shelter belts and windbreaks derives from the reduction in collisions between leaves with the consequent reduction in cuticle damage.

REFERENCES

Armbrust, D.V., Paulsen, G.M. & Ellis, R. (1974). Physiological responses to wind and sandblast-damaged winter wheat plants. Agronomy Journal, **66**: 421-423.

Brun, W.A. (1961). Photosynthesis and transpiration from upper and lower surfaces of intact banana leaves. Plant Physiology, **36**: 399-405.

Campbell, G.S. (1977). An Introduction to Environmental Biophysics. New York: Springer Verlag.

Chafe, S.C. & Wardrop, A.B. (1973). Fine structural observations on the epidermis. II. The cuticle. Planta, **109**: 39-48.

Chambers, T.C. & Possingham, J.V. (1963). Studies of the fine structure of the wax layer of sultana grapes. Australian Journal of Biological Science, **16**: 818-825.

Denna, D.W. (1970). Transpiration and the waxy bloom in *Brassica oleracea* L. Australian Journal of Biological Science, **23**: 27-31.

Dixon, M. & Grace, J. (1984). Effect of wind on the transpiration of young trees. Annals of Botany (in press).

Drake, B.G., Raschke, K. & Salisbury, F.B. (1970). Temperatures and transpiration resistances of *Xanthium* leaves as affected by air temperature, humidity and wind speed. Plant Physiology, **46**: 324-330.

Erner, Y. & Jaffe, M.J. (1983). Thigmomorphogenesis: membrane lipid and protein changes in bean plants as affected by mechanical perturbation and Ethrel. Physiologia Plantarum, **58**: 197-302.

Fryrear, D.W. (1971). Survival and growth of cotton plants damaged by wind blown sand. Agronomy Journal, **63**: 638-642.

Gates, D.M. & Papian, L.E. (1971). Atlas of Energy Budgets of Plant Leaves. New York: Academic Press.

Grace, J. (1974). The effect of wind on grasses. 2. Cuticular and stomatal transpiration. Journal of Experimental Botany, **25**: 542-551.

Grace, J. (1983). Plant-atmosphere Relationships. London: Chapman Hall.

Grace, J. & Russell, G. (1982). The effect of wind and a reduced supply of water on the growth and water relations of *Festuca arundinacea* Schreb. Annals of Botany, **49**: 217-225.

Hall, D.M. (1966). A study of the surface wax deposits on apple fruit. Australian Journal of Biological Science, **19**: 1017-1025.

Hall, D.M. & Jones, R.L. (1961). Physiological significance of surface wax on leaves. Nature, **191**: 95-96.

Hallam, N.D. (1967). An electron mocroscope study of the leaf waxes of the genus *Eucalyptus*, L'Heritier. Ph.D. Thesis, Melbourne University.

Heath, O.V.S. (1938). An experimental investigation of the mechanism of stomatal movement, with some preliminary observations upon the response of the guard cells to 'shock'. New Phytologist, **37**: 385-395.

Hygen, G. (1951). Studies in plant transpiration, I. Physiologia Plantarum, **4**: 57-183.

Jones, H.G. (1983). Plants and Microclimate. Cambridge: Cambridge University Press.

Juniper, B.E. & Jeffree, C.E. (1983). Plant Surfaces. London: Edward Arnold.

Larson, P.R. (1965). Stem form of young *Larix* as influenced by wind and pruning. Forest Science, **11**: 412-424.

Linacre, E.T. (1964). Calculations of the transpiration rate and temperature of a leaf. Archiv für Meteorologie, Geophysik und Bioklimatologie, **13**: 391-399.

Lipetz, J. (1970). Wound healing in higher plants. International Review of Cytology, **27**: 1-28.

Martin, J.T. & Juniper, B.E. (1970). The Cuticle of Plants. London: Edward Arnold.

MacKerron, D.K.L. (1976). Wind damage to the surface of strawberry leaves. Annals of Botany, **40**: 351-354.

Merida, T., Schonherr, J. & Schmidt, H.W. (1981). Fine structure of plant cuticles in relation to water permeability: The fine structure of the cuticle of *Clivia miniata* Reg. leaves. Planta, **152**: 259-267.

Monteith, J.L. (1965). Evaporation and Environment. In The State and Movement of Water in Living Organisms, G.E. Fogg (Ed.) pp 205-235. Cambridge: Cambridge University Press.

Neel, P.L. & Harris, R.W. (1971). Motion-reduced inhibition of elongation and induction of dormancy in *Liquidambar*. Science, **173**: 58-59.

Pieniazec, S.A. (1944). Physical characters of the skin in relation to apple fruit transpiration. Plant Physiology, **19**: 529-536.

Pitcairn, C.E.R. & Grace, J. (1982). The effect of wind and a reduced supply of phosphorus and nitrogen on the growth and water relations of *Festuca arundinacea* Schreb. Annals of Botany, **49**: 649-660.

Raschke, K. (1970). Leaf hydraulic system: rapid epidermal and stomatal responses to changes in water supply. Science, **167**: 189-191.

Roelofsen, P.A. (1952). The submicroscopic structure of cell walls. Acta Botanica Neerlandica, **1**: 99-144.

Schonherr, J. (1976). Water permeability of isolated cuticular membranes: the effect of cuticular waxes on diffusion of water. Planta, **131**: 159-164.

Schonherr, J. (1982). Resistances of plant surfaces to water loss. In Physiological Plant Ecology II Water Relations and Carbon Assimilation. O.L. Lange, P.S. Nobel, C.B. Osmond & H. Ziegler, (Eds). pp 153-179. Encyclopaedia of Plant Physiology 12B. New York: Springer.

Stalfelt, M.G. (1955). The stomata as a hydrophotic regulator of the water deficit of the plant. Physiologia Plantarum, **8**: 572-593.

Taylor, S.E. (1975). Optimal leaf form. In Perspectives of Biophysical Ecology, D.M. Gates (Ed) pp 73-86. New York: Springer-Verlag.

Thomson, J.R. (1974). The effect of wind on grasses II. Mechanical damage in *Festuca arundinacea* Schreb. Journal of Experimental Botany, **25**: 965-972.

Wattendorff, J. & Holloway, P.J. (1980). Studies on the ultrastructure and histochemistry of plant cuticles: The cuticular membrane of *Agave americana* L. *in situ*. Annals of Botany, **46**: 13-28.

Weatherley, P.E. (1965). The state and movement of water in the leaf. In The State and Movement of Water in Living Organisms, G.E. Fogg (Ed) pp 157-184. Cambridge: Cambridge University Press.

Wilson, J. (1978). Some physiological responses of *Acer pseudoplatanus* L. to wind at different levels of soil water, and the anatomical features of abrasive leaf damage. Ph.D. Thesis, University of Edinburgh.

Wilson, J. (in press). Microscopic features of wind damage to leaves of *Acer pseudoplatanus* L. Annals of Botany.

Yamaoka, Y. (1958). Total transpiration from a forest. Transactions of the American Geophysical Union, **39**: 266-272.

Progress in Biometeorology, 1985, Vol. 2, pp. 127-144

SOME EFFECTS OF SHELTER ON THE YIELD AND WATER-USE OF TEA

M.K.V. Carr
Silsoe College,
(Cranfield Institute of Technology),
Silsoe,
Bedford, UK. MK45 4DT

INTRODUCTION

In the late 1960's the absence of shade trees in most new areas of tea (*Camellia sinensis* L.) and their widespread removal from existing tea plantations in many parts of East Africa meant that the crop was exposed to much more wind than in the past. To mitigate the assumed adverse effects of wind on the productivity of the tea crop, particularly during the dry seasons, it became fashionable to plant *Hakea saligna* hedges as shelter belts.

Since little was known about the effects of shelter on the yield and water use of tea, and as the shelter belts were expensive to raise, plant and maintain, and also occupied land which could otherwise be planted to tea, an experimental study was initiated by the Tea Research Institute of East Africa in 1968 to try and quantify these responses (Carr, 1970; 1971). Contrary to expectation it appeared that during extended periods of dry weather, wind sheltered tea plants experienced greater water stress than similar plants which were exposed to more wind. On theoretical grounds this was physically possible (Monteith, 1965) but there was and is still little evidence from field observations to support these theories (Grace, 1977).

In this paper the original work is reviewed and an attempt made to interpret the data, and to explain the mechanisms responsible for the observed effects, using the results of work reported since by other people.

SITE AND METHODOLOGY

The study formed part of a programme of work to investigate the irrigation and water requirements of the tea crop. This was centred in Mufindi at the Ngwazi sub-station of the Tea Research Institute of East Africa in the southern highlands of Tanzania. The shelter experiment was located on the nearby Kilima estate (8 ° 36′S; 35 ° 21′E; 1900 m altitude), owned by Brooke Bond (Tanzania) Ltd., where there were blocks of young clonal tea surrounded on all sides by 4-6 m tall *Hakea saligna* hedges (Fig. 1). The wind blew from the north-east/south-east quadrant for most of the year, changing direction only for a few days (up to 30) during the rainy season when intermittently it blew from the north to west quadrant.

The site was close to a 600 m deep escarpment; the total annual rainfall was high (about 1500 mm) but still seasonally distributed with most of the rain (90%) falling during the six months December to May. This main growing period was followed by cool and dry, often misty weather from June to September, when rates of shoot growth were low. From September to the start of the rains in late November or early December it was warm but shortages of water in the soil reduced yields in the absence of irrigation (Carr, 1974). Representative weather data are shown in Table 1. Wind speeds were not excessive, even on the relatively exposed meteorological site. There was little or no advection during the dry season because of the rapid cooling that occurred as the warm air rose up the escarpment which

Fig. 1 One block of the experimental area photographed towards the end of the 1969 dry season. Note (i) the general ecology of the area with the pockets of rain-forest (ii) the height and porosity of the *Hakea saligna* wind breaks, the prevailing wind blew from left to right (iii) the scorched and droughted tea (clone 230) within those double-row plots closest to the hedge (background) and (iv) the lack of drought symptoms in tea plants pruned at the start of the dry weather (foreground).

was in the direction of the prevailing wind. The soil was deep (greater than 3 m) and well drained. It displayed properties in the field similar to those of a sandy clay loam. Fertilizer was applied at annual rates of 100-150 kg N ha^{-1}, 30-40 kg P_2O_5 ha^{-1}, and 30-40 kg K_2O ha^{-1}.

The useful product of the tea crop is the young shoot, normally two leaves and the unopened terminal bud. These shoots are removed by hand at frequent intervals (every 7-28 days depending on the growing conditions) when they reach harvestable size. Tea is therefore one of the few crops from which regular successive harvests of the useful product can be made from the same plants, a factor which should make it much easier to detect and interpret responses to treatments than for single harvest crops. Every 3 or 4 years most of the foliage is removed in the act of pruning which reduces the height of a crop from about 1.2 to 0.4 m. This stimulates the production of new shoots and makes harvesting easier.

In this experiment the fresh weight of young shoots, harvested at appropriate time intervals throughout the year, was recorded at successive distances from the *Hakea saligna* hedges planted across the direction of the prevailing wind. In total six different clones were represented which, with the exception of one (labelled 207), were all large leaved 'Assam type' in appearance. These had been planted in either 1965 or 1966, in 12 0.4 ha individually sheltered blocks as rooted, container grown plants. The fresh weight of shoots harvested from the centre 40 bushes (spaced 1.2 x 0.9 m) in each of two adjacent rows parallel to the hedge were weighed after each harvest (Fig. 1). With 36 rows (= 44 m) between adjacent hedges this meant that there were 18 plots, 86 m^2 in area, in each block of tea. All the tea plants were therefore within 8-10 times the height of the Hakea hedge and, even excluding any 'back' shelter effects, were sheltered from wind to some extent. In order to help to identify the extent of competition the yields from individual rows adjacent to the hedges were sometimes recorded.

When the experiment began in April 1968 (Phase 1) the tea was unirrigated, but sprinkler irrigation was introduced in June 1971 (Phase 2). The experiment continued until November 1974 when yield recording ceased.

Subsiduary measurements were made during the early part of the experiment to assess the effects of shelter on (a) crop water use, using the gravimetric sampling method to measure changes in soil water content, (b) plant water status, using the pressure chamber technique to estimate shoot water potentials and the infiltration method to assess the degree of stomatal opening and (c) the microclimate, using dish evaporimeters to estimate potential evaporation rates, sensitive cup anemometers to measure windspeed and an Assman psychrometer to assess air temperature and humidity just above the surface of the crop. Associated measurements of the effect of shelter on some of these variables were also made at the Ngwazi sub-station and later in Kericho, Kenya. Details of the techniques used can be found elsewhere (Carr, 1971).

RESULTS

Yield responses

The experiment was not statistically designed; it was imposed on an existing area of tea. Since there were 2 blocks of most clones there was limited replication, but no randomisation. However, a factorial analysis of the effects of (i) season (ii) distance from hedge (iii) clone and (iv) block location on most of the yield data for 4 of the clones in the period April 1968 to June 1971 was possible but, because of large seasonal effects on production, the results for each season were analysed separately.

The results obtained were not easy to interpret; responses to shelter by individual clones were influenced in part by differences in their sensitivity to (a) drought and (b) a stem canker *Phomopsis theae*, and planting date,

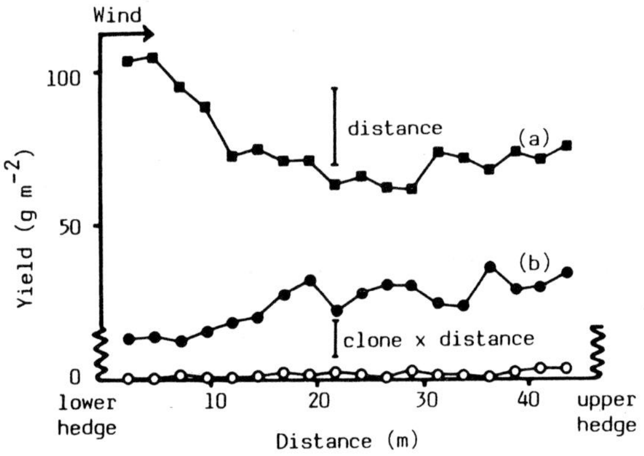

Fig. 2 Yield of fresh shoots in relation to distance from lower hedge.
(a) Mean yields from 3 clones (50, 210 and 230) April to September 1968, end of rains through cool, dry season.
(b) Mean yields from clones 207 and 230 (●), 60% ground cover and clones 50, 210 (○), 100 % ground cover during warm, dry months of October and November, 1968.
Bars represent least significant difference at $P = 0.05$ for appropriate variables.

and hence degree of ground cover early in the experiment.

Phase 1; rainfed

Fig. 2(a) shows the mean yield data for the 6 months (April – September) from the start of the experiment in 1968 for 3 clones. During this period of cool, mainly dry weather after the end of the main rainy season, yields were low but there was evidence of the beneficial effects of shelter. By contrast in October and November, when it was warmer but still dry, yields from 2 clones (207 and 230) increased as exposure to wind increased (Fig. 2b). These clones had been planted a year later than the other two (50 and 210) and the canopy still only covered about 60% of the ground surface compared with 100% for the 2 older clones. Yields from these clones (50 and 210) were very small everywhere at this time because of water stress resulting from greater rates of water use earlier in the dry season.

Shelter was again beneficial over the following 6 month period which included the rains and the early part of the cool, dry season (Fig. 3a) but, as in the previous year, yields in the warm, dry months October and November 1969 again increased with distance away from the hedges (Fig. 3b). By the end of the dry weather severe scorching and defoliation was clearly visible especially in the 10-15 rows in the immediate lee of the shelter belts. Clone 230 was particularly badly affected in this way with the result

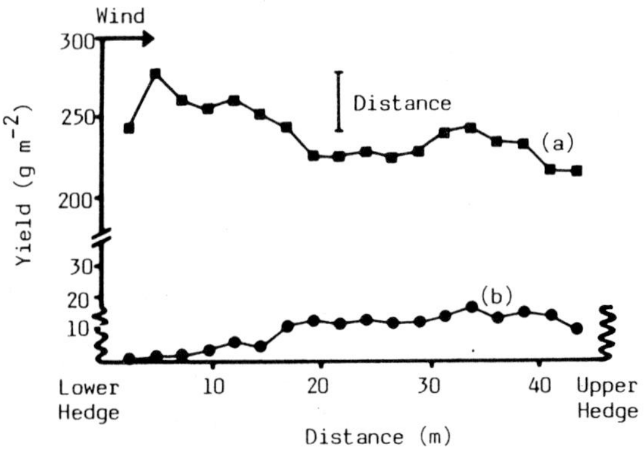

Fig. 3 Yield of fresh shoots in relation to distance from lower hedge.
(a) Mean yields from 4 clones (50, 207, 210 and 230) January to September, 1969, rains and cool dry season. Clones 50 and 210 were pruned in June.
(b) Mean yields from 2 clones (207 and 230) October and November, 1969, dry season. Clones 50 and 210 had been pruned at start of dry weather.

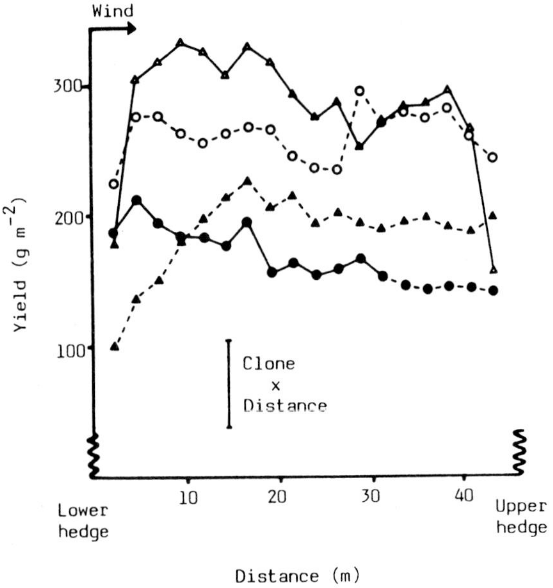

Fig. 4 Yield of fresh shoots in relation to distance from lower hedge for each clone, January to August 1970, rainy season and cool dry weather. Clones 50 (○) and 210 (△) had both been pruned at the start of the preceding dry season whilst clones 207 (●) and particularly clone 230 (▲) had suffered severe water stress over the same period. Clones 207 and 230 were both pruned in June 1970.

that it took a long time for these bushes to refoliate and to come back into production once the rains began especially in the lower, more sheltered block which suffered most (see Fig. 1). This was reflected in the relative yields recorded during the following rainy season (Fig. 4). By comparison the drought tolerant, but general low yielding 'China type' clone 207, recovered quickly and there was again a net benefit from shelter at this time.

Both clones 50 and 210 had been pruned at the start of the dry weather, an act which conserves soil water by restricting transpiration. Levels of water stress were therefore relatively low until refoliation occurred towards the end of the dry season (Carr, 1971). As a result clone 210 in particular benefited from shelter during the following rainy season (Fig. 4) but the response of clone 50 was influenced in part by the affect of shelter on the incidence of a stem canker, *Phomopsis theae*, during the dry weather (Fig. 5). This disease affects susceptible clones (such as clone 50) when the soil is dry (Shanmuganathan & Rodrigo, 1967). Once the rains began the bushes

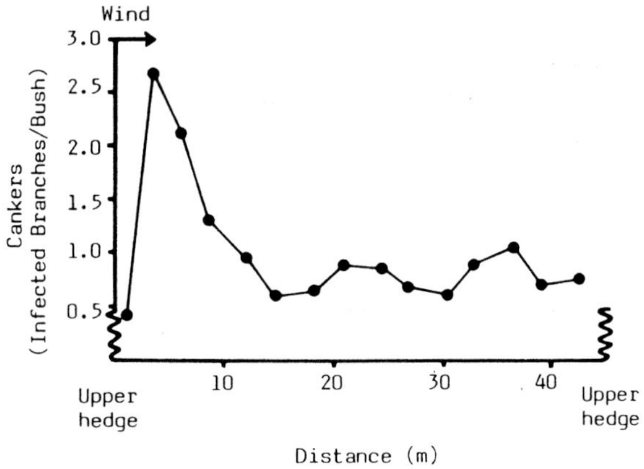

Fig. 5 The incidence of *Phomopsis theae* in relation to distance from *Hakea saligna* trees on clone 50 as observed towards the end of the 1969 dry season. An 'infected branch' was defined here as one with a canker greater than 20 mm in length.

which were badly infected took longer to come back into full production than others. These differential responses between clones to shelter during the 1970 rainy season, due it appears to a variety of reasons, were statistically highly significant. In July 1970, at the start of the dry season clones 207 and 230 were both pruned and regular recording of yield did not recommence until the beginning of the next rainy season. Yields for both clones left unpruned were very low during the dry season but, as in 1968 and 1969, there was visible evidence that bushes at distance from the shelter suffered less in the dry weather than those in the immediate lee of the hedge. Clone 50 was again severely infected by *Phomopsis theae*, particularly close to the shelter, and this had a marked affect on rates of recovery and yield distribution during the following rainy season with yields increasing with distance from shelter in contrast to the other 3 clones.

Fig. 6 shows the cumulative yield totals for the 27 month period April 1968 to June 1971 averaged for all 4 clones. The effects of sheltering on yield were generally small, being masked by the differential responses by clones and the effects reported above. For clones 207 and 210, which were both least affected by drought, the overall yield increase for the 10 most sheltered rows compared with the 10 most exposed (in both cases excluding the 2 rows adjacent to the hedges) was 20%, compared with 5% for clone 230 and an 11% reduction in the disease affected clone 50.

At the end of the first phase of the experiment, when there was no

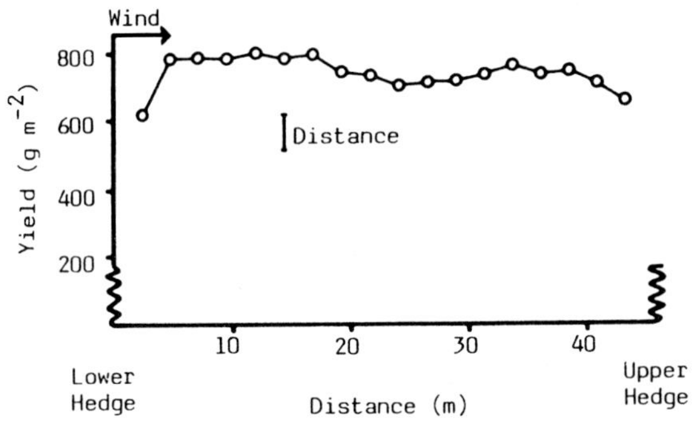

Fig. 6 Mean yield of fresh shoots for all 4 clones in relation to distance from lower hedge for the whole of Phase 1 of the experiment. April 1968 to June 1971. 800 g m^{-2} of fresh shoots is equivalent to about 1600 kg ha^{-1} of made tea.

irrigation, these were the principal conclusions.

1. During the main growing season and the cool dry months which followed, shelter had a beneficial effect on rates of shoot growth and yield. Rates of refoliation after pruning were faster during the dry season in sheltered areas.
2. Towards the end of the extended dry season when it was warm, shelter had an adverse effect on production, the degree of yield reduction being influenced by the stage of development of the plants and their relative susceptability to drought and infection by *Phomopsis theae*.
3. If the adverse effects were severe there was a carry-over effect into the next rainy season which resulted in lower yields close to hedges until recovery was complete.
4. Apart from the rows immediately adjacent to the hedges there were no obvious effects of competition between the *Hakea saligna* hedges and the tea plants.

Phase 2: Irrigated

In 1971 irrigation was introduced and during this and subsequent years, at the discretion of the estate management, 200-400 mm of water were applied in 2 to 4 applications of 100 mm (net) at approximately monthly intervals from August onwards. Crops were not irrigated in the year of pruning. Alternate rows of shelter belts were also uprooted by the estate in 1971 so the yields reported refer to one sheltered block only with the 'upper' hedge removed and no formal statistical analysis was possible.

The mean yields for all 4 clones recorded over the 21 months January 1973 to September 1974 are shown in Fig. 7a. This period includes the 1973 dry season at the start of which all the plants were pruned, regular harvesting only recommencing in the following January. Yields within about 20 m (3*H*) of the shelter belt were greater than those elsewhere in the field. No such effect was observed in the blocks where the hedges had been removed.

The corresponding yields during the warm, dry months of October and November 1974, when a total of 300 mm or irrigation was applied in August, September and November are shown in Fig. 7b. With only about 60 mm of rain during the period June to October irrigation increased yields substantially compared with previous dry seasons (see Fig. 2 and 3). In

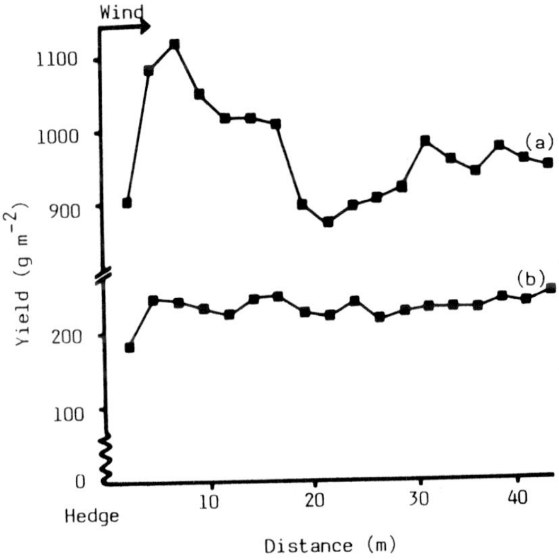

Fig. 7 Yield of fresh shoots in relation to distance from shelter belt.
(a) Mean yields from 4 clones (50, 207, 210 and 230) January 1973 to September 1974, rains and cool seasons; plants pruned 1973 dry season.
(b) Mean yields from 3 clones (207, 210 and 230) October and November 1974, warm dry weather but irrigated. Results for one block of each clone only in both cases.

contrast to the years before irrigation was introduced there was no evidence, for 3 of the clones, of any adverse effects of shelter on yields at this time. Clone 50 was though an exception with yields increasing over the first 16 m away from the shelter, probably as a result of the effects of *Phomopsis theae*.

Effects of shelter

Microclimate

Measurements showed that the daytime wind speed, recorded over intervals of about 30 minutes 0.15 m above the crop canopy, averaged about 0.5 m s^{-1} in the lee of the shelter increasing to about 2 m s^{-1} in the most exposed areas. The daytime air temperatures immediately above the canopy were up to 1-2 °C higher (usual range 17-19 °C) in the lee of the shelter than in the more exposed areas. The saturation vapour pressure deficit (svpd) was always low, being in the range 0.5-0.8 kPa, and shelter only increased it slightly (ca. + 0.06 kPa). In 186 observations taken during 1968 and 1969 the maximum air temperature and svpd recorded in the field were 25 °C and 1.7 kPa. Shelter reduced evaporation (from Hudson dish evaporimeters) for a distance of up to 4 times the height of the shelter belt, which was less in extent than the corresponding reduction in windspeed (Fig. 8).

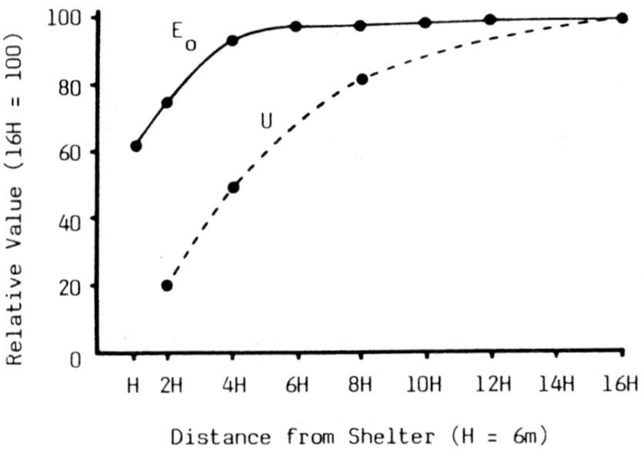

Fig. 8 **Effects of *Hakea saligna* hedges on evaporation from Hudson dish evaporimeters (E_o) and windspeed (*u*) measured above a canopy of tea.**

Crop water use

There was circumstantial and other evidence to suggest that during the extended dry season wind-sheltered tea came under greater water stress than wind-exposed tea. This effect was mitigated if the tea plants were pruned during the early part of the dry season. Recovery from pruning was always more rapid in the wind-sheltered areas, since the act of pruning conserved water (Carr, 1971). Fig. 9 shows the changes in soil moisture content during the 1968 and 1969 dry seasons, averaged over the entire 1.5 m depth of soil, at two positions relative to the hedges, for unpruned crops. In both years there were strong indications that more water (20-30 mm m^{-1}) was available in the wind exposed areas than in the sheltered parts, certainly until the middle of October by which time both profiles were close to permanent wilting point. In 1969 there was evidence of drying to a depth of 2.1 m. The corresponding maximum actual soil water deficits were about 260 mm in 1968 and 320 mm in 1969. For comparison the estimated potential soil water deficits ($E_t = 0.85E_o$) by the end of October were about 420 mm in 1968 and 450 m in 1969.

Measurements of shoot water potentials showed clearly how unpruned, wind-exposed tea experienced greater water stress than wind-sheltered tea by the end of the dry season (Carr, 1971) whilst estimates of stomatal opening using the infiltration technique made throughout one dry season showed a similar response as the dry season progressed (Fig. 10). In June and July when soil water was still freely available, there was some evidence that the stomata may have been wider open in the sheltered areas than in

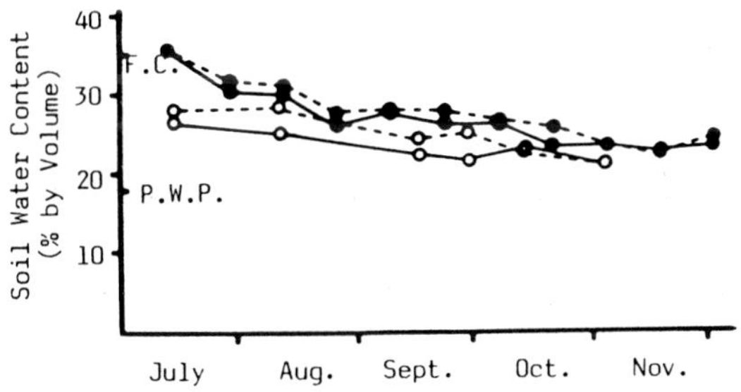

Fig. 9 Changes in soil water content during the 1968 (•) and 1969 (O) dry seasons at row 7 (– – – –, 8-9 m from the hedge) and row 28 (—— ——, 34 m). Each point represents the mean of 40-80 individual assessments of soil water content, duplicate samples having been taken at 0.3 depth increments down to 1.5 m from 6 or 8 (1968) and 4 or 6 (1969) profiles at each position. From the end of July to mid-November only about 10 mm of rainfall was recorded in 1968 and 40 mm in 1969. The main rains also ended early in 1969. The dry bulk density has been taken to be 1.3 g cm^{-3}. Soil moisture contents corresponding approximately to field capacity (F.C.) and permanent wilting point (P.W.P.) based on field measurements are also indicated.

exposed rows. Afterwards though the situation appeared to reverse with the stomata of wind-exposed plants wider open than those of similar sheltered plants.

Mechanisms

The results suggest that shelter is beneficial to tea production in Southern

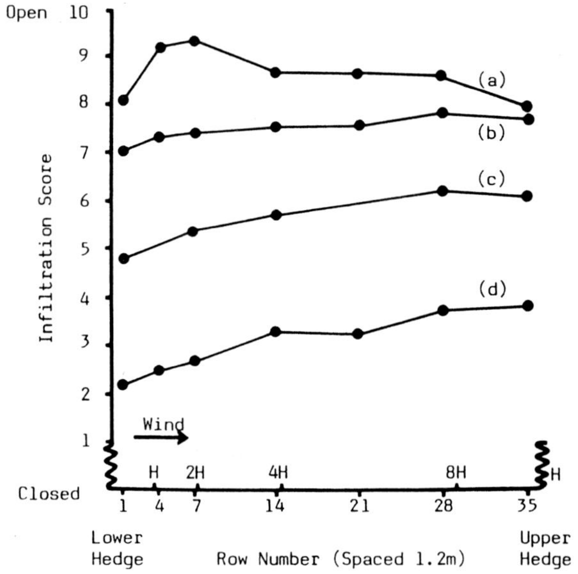

Fig. 10 Changes in the degree of stomatal opening (abaxial surface), as indicated by the infiltration technique, with distance from *Hakea saligna* hedges during the 1968 dry season. All values are the means for 3 clones (50, 210 and 230) with 5 measurements on each clone, at each position and on each date.
(a) At start of dry season; readings taken on 15th and 29th July with estimated potential soil water deficit (S.W.D.) 50-100 mm
(b) 12th August; S.W.D. 140 mm.
(c) 26th August and 9th September; S.W.D. 180-230 mm.
(d) End of dry season and beginning of rains 4th and 18th November and 2nd December; S.W.D. 430-480 mm.

Tanzania during the rainy and cool, dry seasons but deleterious during warm, dry weather unless there is irrigation. Possible mechanisms responsible for these responses are discussed below.

Temperature

The yield of tea is a function of shoot number, rates of shoot extension and shoot weight and Tanton (1982) has recently showed that seasonal effects on production in Malawi are mainly the result of differences in rates of shoot-growth as affected by air temperature:

$$Y = 0.063t - 0.804$$

where Y is the rélative growth rate of shoots in cm cm^{-1} $week^{-1}$ and t is the effective weekly mean air temperature (°C) above a base of 12.5 °C. Providing shoot growth was not limited by excessively large saturated vapour pressure deficits (>2.3 kPa) Tanton (1982) showed that about 500 day °C above a base of 12.5 °C were needed for shoots to reach a harvestable length of 150 mm after the release from apical dominance.

Mean air temperatures in Mufindi were always a long way below the optimum for tea and in the 'winter' months of June, July and August were only just above the base temperature for growth (Table 1). Using the relationships developed by Tanton (1982) Fig. 11 shows the effect of a 1 °C increase in the mean air temperature on the likely time taken for shoots to reach harvestable size. Throughout most of the main growing season in Mufindi, when daily mean air temperatures were in the range of 17-18 °C, a 1 °C rise resulting from shelter represented an increase of about 20% in the rate of shoot extension (or yield). In the winter months when mean air temperatures were only 12-13 °C the same temperature rise corresponded to a doubling of the yield. These values agree closely with the observed yield

Table 1. *Representative weather data for Kilima Estate, Mufindi, Tanzania. (8°6'S; 35°1'E; 1900 m altitude) in 1968 and 1969)*

Month	Air Max.	Air Min.	Temp (°) Mean	Dew Point°C	Wind (km d^{-1})	R_s** (cal $cm^{-2}d^{-1}$)	E_0 (mm)	Rain (mm) (10 year mean)
J	22	14	18	16	160	400	130	237
F	22	14	18	15	180	425	130	196
M	21	13	17	15	170	405	130	266
A	20	13	16.5	15	180	370	110	358
M	18	11	14.5	13	260	325	100	130
J	16	10	13	11	230	360	95	36
J	15	9	12	9	260	390	100	29
A	16	10	13	9	270	360	110	12
S	18	11	14.5	12	270	470	130	8
O	21	12	16.5	13	280	510	155	15
N	21	13	17	14	250	450	150	57
D	22	14	18	15	180	450	140	153
TOTAL	–	–	–	–	–	–	1480	1517

* Mean 0900 and 1500h ** Solar radiation

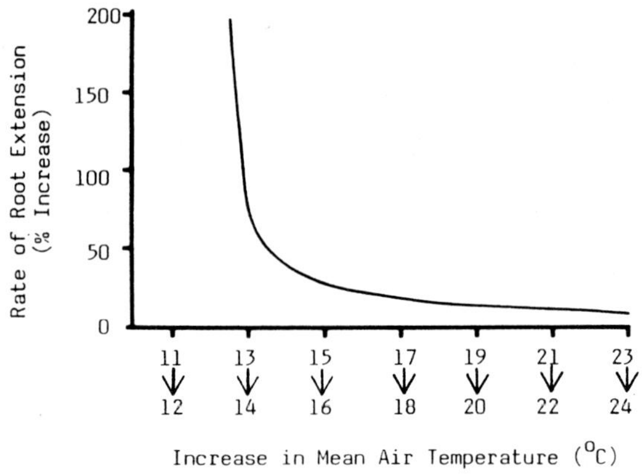

Fig. 11 The effect of a 1 °C rise in mean air temperature on the rate of extension of tea shoots (after Tanton, 1982).

differences in this experiment and by estate staff elsewhere when water was not limiting. One implication of this is that in tea areas where mean air temperatures are much higher than in Mufindi the benefits from shelter are likely to be much less than these reported here.

Water use

The expected advantages from sheltering in the dry season did not occur. Previously wind-induced stomatal closure has been implicated by others to explain observations that plants exposed to wind may sometimes use less water than those which are sheltered (e.g. Stalfelt, 1955; Winter, 1965) and there was some evidence in this experiment that, when water was freely available, stomata were wider open close to the shelter (Fig. 10). Supporting evidence for this was also obtained at the nearby Ngwazi sub-station and later at Kericho in Kenya on plants well supplied with water (Fig. 12). Ignoring the 3 points at very low wind speeds, where clearly the large aerodynamic resistance controlled transpiration rates (see below) an increase in wind speed from 0.5 to 2.5 m s^{-1}, recorded 0.15 m above crop surface, corresponded to a 20% reduction in the degree of stomatal opening.

By implication these data suggest that, during the main growing season and the early part of the dry weather, canopy resistance (r_c) could have been less in the lee of the shelter than in wind-exposed rows. Similarly, since the aerodynamic resistance (r_a) is inversely proportional to windspeed, (for the

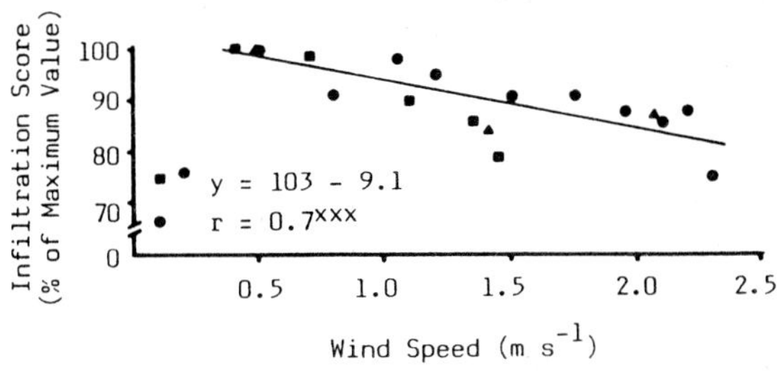

Fig. 12 Relation between degree of stomatal opening (infiltration score) and windspeed measured 0.15 m above crop surface. The results were obtained under the following conditions:

■ Clone number 1, midday estimates on 7 occasions during rainy season and irrigated cool dry season in 1969 at 6 positions in relation to *Hakea saligna* hedge at Ngwazi sub-station, Tanzania. Each point is the mean of 65 measurements.

O As above but readings taken on 5 occasions during the warm dry season in 1969 with irrigation at 12 positions. Each point is mean of 50 measurements.

▲ Tea plants grown from seed, readings taken on 5 days in December 1970 and January 1971 when soil close to field capacity in Kericho, Kenya. Diurnal changes observed by taking 6 or 7 separate readings on each day at 3 positions relative to shelter hedge. Each point is the mean of 330 measurements.

— The linear regression is based on the mean values only (n = 18) after ignoring the 3 anomalous points at low windspeeds. Interpolated windspeed values were used where necessary.

conditions described the roughness coefficient z_o can be assumed to be constant) the value of r_a should be greater in the sheltered areas than elsewhere.

In a detailed energy balance study on mature, estate grown seedling tea in Kenya, Callander and Woodhead (1981) estimated the canopy resistances in the rainy season to be of the order of 80 s m^{-1}. They decreased with increasing irradiance and increased with the saturated vapour pressure deficit of the air. Depending on the weather conditions estimated values of r_a ranged from 30 s m^{-1} to more than 150 s m^{-1} (see also Squire & Callander, 1981). The corresponding value for the aerodynamic resistance (r_a) between the crop and 2 m above the canopy were found by Callander and Woodhead (1981) to be typicaly 20 s m^{-1} for an unsheltered crop (height, 1.2 m; z_o, 0.20 ± 0.07 m; zero plane of displacement d, 0.70 ± 0.1 m).

According to the analysis of Thom and Oliver (1977) the value of the ratio

Table 2. *Effects of changes in the ratio (n) of the canopy (r_c) and aerodynamic (r_a) resistances ($s\,m^{-1}$) on the estimated value of the crop factor E/E_o (see text for details of assumptions made)*

	Wind sheltered				Wind exposed			
	r_c	r_a	n	E/E_o	r_c	r_a	n	E/E_o
Rainy season	60	50	1.2	1	80	20	4	0.7
Late dry season	200	50	4	0.7	160	20	8	0.5

between actual evapotranspiration (E) and open water evaporation (E_o) for a given crop depends on the ratio of the canopy (r_c) and aerodynamic (r_a) resistances to vapour exchange. Squire and Callander (1981) have plotted the relation between the crop factor E/E_o and n for the tea crop. Using this relationship it is possible to estimate the effect of changes in the ratio r_c/r_a on the crop factor E/E_o as affected by windspeed and the water status of the crop (Table 2).

In deriving the crop factor various assumptions have been made about r_c and r_a. The basic values for wind exposed, well watered plants are the same as the 'typical' ones cited by Callander and Woodhead (1981). For well watered, sheltered plants r_c is assumed to be less than this (stomata wider open) and r_a higher (less wind). As the crop comes under water stress during the dry season r_c increases, especially in the sheltered areas whilst the corresponding values of r_a remain the same.

The values of the E/E_o ratio so obtained support the field observations and empirical evidence that rates of water use of wind-sheltered tea can be greater than those of wind-exposed tea. The results are also consistent with those predicted by Grace (1977) for similar climatic and crop conditions.

CONCLUSIONS

These contrasting responses of tea to shelter demonstrate clearly the general problem of trying to interpret the results of agricultural shelter field experiments when so many interacting soil, plant and climatic factors can influence crop responses. Nevertheless despite the complications it is possible to suggest logical explanations based on sound physical principles for the observed effects on yields. The benefits of shelter during the main growing season and the cool dry weather appear to be due to associated change in air temperatures whilst the adverse effects of shelter during the extended dry season, in the absence of irrigation, can be explained by the relative changes in the ratio of the canopy and aerodynamic resistances on actual evapotranspiration rates.

Whatever the mechanisms for these responses the problem still remains though of deciding what advice should be given to tea producers. Under conditions similar to those in the Mufindi district of Tanzania, it appears that shelter is beneficial when there is enough rain or when the crop is irrigated. Without irrigation there may or may not be a net benefit depending on the stage of development and age of the crop, particularly as it affects the shoot:root ratio and hence the susceptibility of the crop to drought. In tea areas where the mean air temperature is above about 16-17 °C shelter belts are unlikely to be worthwhile unless the cooling effect of wind is large or there is a strong advection.

The extent of the dilemma is illustrated by a letter written by T.C.E. Congdon, Estates Director of Brooke Bond (Tanzania) Ltd. in March 1983:

> "So for the past 10 years we have been solemnly digging out the *Hakea* we spent the previous 10 years planting . . . "

Since irrigation has also been introduced during the last 10 years it is debatable whether this was the correct course of action but I leave you, the reader, to decide.

ACKNOWLEDGEMENTS

I thank Colin Congdon and Brooke Bond (Tanzania) Ltd. for their help and co-operation with the field studies. The work was funded in part by the UK Overseas Development Administration through a grant to the former Tea Research Institute of East Africa. Staff of this Institute, in particular Francis Shirima, Boniface Miho, and Benedict Kiduko, assisted with the data recording. I thank them, and also Alan Clewer of Wye College (University of London) for his valuable help with the statistical analyses.

REFERENCES

Callander, B.A. & Woodhead, T. (1981). Canopy conductance of estate tea in Kenya. Agricultural Meteorology **23**: 151-167.

Carr, M.K.V. (1970). The role of water in the growth of the tea crop. **In** Physiology of Tree Crops (eds. L.C. Luckwill & C.V. Cutting). pp. 287-305, Academic Press, London.

Carr, M.K.V. (1971). The internal water status of the tea plant (*Camellia sinensis*): some results illustrating the use of the pressure chamber technique. Agricultural Meteorology **9**: 447-460.

Carr, M.K.V. (1974). Irrigating seedling tea in southern Tanzania: effects on total yields, distribution of yield and water use. Journal of Agricultural Science, Cambridge **83**: 363-378.

Grace, J. (1977). Plant Response to Wind. pp. 99-103, Academic Press, London.

Monteith, J.L. (1965). Evaporation and environment. Symposium of the

Society for Experimental Botany **19**: 205-234.

Shanmuganathan, N. & Ridrigo, W.R.F. (1967). Studies on collar and branch canker of young tea (*Phomopsis theae* Petch) II Influence of soil moisture on the disease. Tea Quarterly **38**: 320-330.

Squire, G.R. & Callander, B.A. (1981). Tea Plantations. Chapter 7 **In** Water Deficits and Plant Growth Vol. VI (ed. T.T. Kozlowski). pp. 471-510, Academic Press, New York.

Stalfelt, M.G. (1955). The stomata as a hydrophytic regulator of water deficit of the plant. Physiologia Plantarum **8**: 572-593.

Tanton, T.W. (1982). Environmental factors affecting the yield of tea (*Camellia sinensis*). I. Effects of air temperature. Experimental Agriculture **18**: 47-52.

Thom, A.S. & Oliver, H.R. (1977). On Penman's equation for estimating regional evaporation. Quarterly Journal of the Royal Meteorological Society **103**: 345-357.

Winter, E.J. (1965). Some effects of wind upon vegetable crop plants. Scientific Horticulture **17**: 53-60.

Progress in Biometeorology, 1985, Vol. 2, pp. 145-154

WIND PROTECTION IN TRADITIONAL MICROCLIMATE MANAGEMENT AND MANIPULATION – EXAMPLES FROM EAST AFRICA

C.J. Stigter
Section of Agricultural Physics,
Physics Department,
University of Dar es Salaam,
Dar es Salaam, P.O. Box 35063, Tanzania.

INTRODUCTION

Traditional farmers have developed many skills in managing and manipulating soil and crop microclimate (Wilken, 1972). With the rising economic problems almost everywhere in the Third World it becomes more and more pressing to explore further the cheapest means of reducing yield fluctuations and increasing production. As farmers often make diligent use of locally available materials and existing ecological conditions, it is worthwhile to study first their methods of improving crop growth conditions before proposing any changes in existing practices (Stigter, 1983).

In an earlier review on traditional applications of mulches (Stigter, 1984b) examples were given of the use that is made of their protective properties. These included shading and reducing adverse effects from mechanical impacts of rain, hail and wind, such as soil erosion and damage of aerial plant parts. Protection of crops from mechanical wind effects was also mentioned in combination with radiation protection in another earlier paper reviewing traditional shade management and solar radiation manipulation (Stigter, 1984a). Natural mulch and artificial mulch with non-transparent materials have a definite shading effect. However, open shades do not have an additional mulch character (Stigter *et al*, 1984), although in many cases they have a reducing effect on average wind speed near the surface. This is understandable when we realize that horizontal air movement in the atmosphere is reduced by friction with the earth's surface.

Close to the surface, average horizontal wind speed reduces quickly with height. But even in erosion of flat bare soils, air movement due to topographical features and turbulent air movement – the eddies which make up wind structure (e.g. Grace, 1977) – are as important as agents in wind stress as the drag forces of horizontal air movement. Apart from thermal genesis, such turbulence is a direct consequence of the "roughness" of the terrain, wind meeting natural or man-made obstacles. Type – that is morphology, flexibility, composition or construction – and spacing of such obstacles (roughness elements) determine the character and evolution of the resulting local wind field. This is true for what might have to be protected and for what is used as protection. It is here that ingenuity to prevent mechanical damage to soils and plants by farmers' management and manipulation comes in.

In what is called primary wind injury in plant stress typology (Levitt, 1972), damage to whole plants and trees (swaying, shaking, bending, lodging) and plant parts (premature fruit and flower shedding, breakage, bruises, lesions, abrasion) is basically due to the mechanical stress caused by asymmetrical air pressures acting on plant parts. However, on the leaf and root scale secondary injury effects, by rubbing of adjacent leaves and by soil particles or "scouring" by carried particles, are involved as well. Information as to the precise nature of primary and secondary pressure injury or of tolerance of, and adaptation to, such wind action is very incomplete (Sturrock, 1975; Grace, 1977; Jaffe, 1980). Other secondary injury effects of winds are also known. Air movement plays a role in evaporation stress, in carbon dioxide and soil moisture depletion, in cold air drainage and mixing and as a collector and carrier of such agents as soil particles, salt, pathogens and insects. Protection against mechanical stress usually influences damage potential of those secondary factors. In this respect, until a decade ago emphasis in regard to wind protection of plants was mainly put on reduced evaporation stress and improved soil moisture conditions (Van Eimern, 1964; Marshall, 1967; Levitt, 1972). Recent work throws doubts on this simplistic explanation of the causes of shelter effects on crop growth and development (Van Eimern, 1968; Rosenberg, 1974; Sturrock, 1975; Grace, 1977; Jaffe, 1980).

Only a few exceptional but interesting examples exist in which traditional farmers are aware of secondary effects. However, judging from the existing literature, protection against primary wind injury must have been widespread since antiquity. Nevertheless, the line between traditional and modern practice has become vague in many cases (Hurst & Rumney, 1971; Wilken, 1972). The examples described below are largely obtained from the recent public contest on management and manipulation of microclimate by traditional farmers in Tanzania (Stigter, 1982; 1984a; 1984b). They confirm, for East Africa, that wind protection is a long-standing indigenous practice.

SOIL EROSION

The single major reason for the use of windbreaks around the world is to control soil blow and its consequences (Sturrock, 1978). Soil erosion is a cumulative effect by flying particles getting impetus from horizontal winds close to the surface. Travelling up and down in the turbulent wind, soil

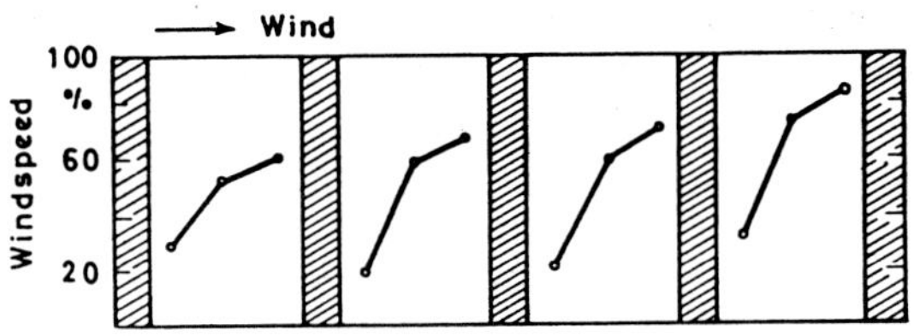

Fig. 1. Wind speed reduction between strips of maize (*Zea mays*). Distances between 3 m wide strips were 12 m. Windbreak height was 1.5 m. From Van Eimern *et al* (1964) after Kaiser.

particles pass on their momentum in meeting and liberating other particles not yet airborne (e.g. Van Eimern, 1964). Managing living and dead vegetative cover is the most effective and practical method for controlling wind erosion (Lyles & Allison, 1981). Roughness of the surface then diminishes soil erosion in two ways. It firstly reduces the horizontal drag on the soil and the flying particles (Fig. 1). Secondly the roughness elements catch the airborne particles not yet redeposited, making them harmless as erosion agents.

In most cases traditional farmers are only aware of wind speed reducing effects. Sometimes they know how to exploit the basic principles skillfully. Tanzanian farmers in some areas use soil ridges on flat land for this purpose. They also employ untilled land strips bearing tall grass or other natural vegetation to serve as wind breaks. As we will illustrate with more examples in the next part of this chapter, farmers are not only very well aware of the fact that shelterbelts or windbreaks in rows orientated perpendicular to the prevailing wind direction influence wind speed (Battan, 1983); but they have also learned that an area is sheltered by growing or leaving elements of the natural vegetation throughout the field. This is also done against erosion in Tanzania. We have not observed the equivalent of the surprising example of Zuni Indians, who improve conditions of eroded soil by setting out rows of low sagebrush on worn out or barren fields to induce deposition of fresh sand and dust (Wilken, 1972).

Traditionally, cover crops were used widely in Africa as they were known to be surprisingly effective in controlling soil blowing. Although introduction of cash crop farming has discouraged their use (de Vos, 1975), in some places traditional farmers still use grass, deeper rooting cover crops or selective weeding for erosion prevention. Recently a new future for cover crops has been suggested (Harrison, 1982). The only documented examples in East African cash cropping are cover crops in sisal rows and nurse crops (often oats) between the rows of young tea (Acland, 1971).

WINDBREAKS

Belts of trees or shrubs and windbreaks (grown or erected) to provide shelter for crops are well known everywhere. Clearings in forests might have been the earliest example (Wilken, 1977). Stone walls and dense hedges at all sides were, and often still are, serving more than one purpose. These may be indicating plot boundaries, providing discouragement to intruders, preventing water run off and protecting crops from strong winds and blown material. It is very likely that this is the main reason that the majority of windbreak examples from many areas in Tanzanian food crop growing are of the "around the field" type. It is less likely that this is purely for wind protection as very strong winds are usually from prevailing directions, with the exception of some periods during thunderstorms.

The examples we collected include the use of raised sides of rice beds, the making of heaps of soil more than 1 m wide and 75 cm high around the shamba (agricultural field), the building of walls of maize stalks around cereal fields, the growing of hedges and trees (for example mangoes) around fields of millet and wheat and the sowing of perennial castor. A rather special example of the use of a windbreak in cash cropping is the growing of protection trees around tree farms to get valuable straight poles in West Kilimanjaro, an area specifically suffering from strong winds (cf. Kangele, 1980). To illustrate again the difficulty of distinguishing between indigenous and imported techniques of microclimate manipulation (Stigter, 1984a) we refer to an example from Uganda (Acland, 1971). A banana variety used there for making beer is often grown around the perimeter of a plot of bananas; it is left unpruned and therefore makes an effective windbreak. Finally, farmers in Arusha region are reported to plant trees to decrease the passive spread of wind borne insects, one of the few examples of awareness of secondary effects.

In East Africa forests and forest strips are used as shelterbelts by growing food crops at their edges and, for example, strips of maize and sorghum to protect other crops in intercropping. However, the majority of windbreaks grown or erected only for the sake of wind protection are found in cash cropping. Acland (1971) mentions that in case of absence of natural shelter there is a need to use windbreaks for nurseries of tobacco and coffee seedlings, for citrus orchards and for tea. Tea is the only crop for which such details as multirow application, necessity of barrier straightness, danger of strong air movement in gaps (funnelling), influence of break permeability on turbulence structure and secondary effects on transpiration and water stress are mentioned. No indication was found that such details are known to traditional farmers in East Africa.

Contrary to this, there is ample evidence that farmers are traditionally very much aware of what was only a beginning point of discussion in the scientific wind shelter literature of the early sixties (Van Eimern, 1964). It was summarized recently as follows. When a regular decrease in wind velocity is required, suitable roughness elements, preferably trees, bushes etc should be spread evenly and in small groups over the country (ILACO, with De Vries & Stigter, 1981). This is the most common way of reducing the wind's power in traditional farming in Tanzania. Bananas are traditionally grown either scattered, but in patches for self protection, or with vegetables in mixed cropping. This is among other things practiced to give shelter to vegetables and bananas through its configuration (cf. Beets, 1982). Introduction of commercial coffee growing has been relatively easy in some parts of East Africa because it could continue the existing farming system of mixed cropping (Hyden, 1980). This included, among many others, bananas traditionally sheltering young coffee seedlings and castor sown among arable crops. Palm-like trees and many other forest trees (some leguminous) are reported to be left scattered in fields with banana and coffee (single and together), maize, tobacco and vegetable crops, all providing wind reduction. Cardamom is grown in the forests for shelter. The interest that is being shown in grazing and arable agroforestry (Sturrock, 1975) must also be partly due to its evident value for the traditional farmer in regard to shading and wind protection (cf. Greenland, 1975; Wilken, 1977).

OTHER PROTECTION

African farmers obviously know the influence of topography on air movement, planting in valleys or at wind-protected sides of hills where this reduces damage. Apart from protection against wind, New Guinean Highland farmers manage the microclimate by growing frost-vulnerable food crops scattered in small patches, promoting the drainage of cold air away from the plants (Waddell, 1983). We did not observe this practice in African Highlands.

Some rather intensive methods of reducing wind damage are found in Tanzanian traditional agriculture because of its small plots and its labour intensity. In some cases, provided that the associated increase in humidity will not encourage pests or diseases, weed residues and soil are put around stems of plants with short roots to improve stability. Comparably, small farmers select the deepest soil for transplanting their banana suckers and tobacco seedlings to provide additional support, guarding them in this way against strong winds. Recent research findings show the wisdom of that practice (Casada *et al*, 1980a; 1980b). It has been reported that farmers smear mud onto plant stems to make them stiff and less easy to be bent by wind blows. This kind of "nursing" resembles an example from Japan (Uemura *et al* 1974) indicating that increase in resistance to water and nutrient transport at breaking and bending portions of rice culms can be reduced "by raising up treatment, which has been practiced for many years by farmers from experience".

Supporting individual trees or even digging individual pits for wind prone trees (banana, paw paw) is also practiced for damage reduction, although in the latter case access to and collection of soil moisture may be the primary aim. Ratooning, the practice of cutting the main stems to stimulate

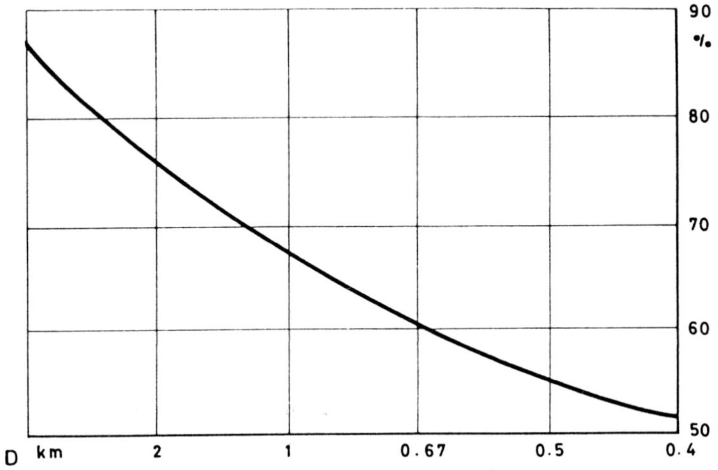

Fig. 2. Effects of forest-shelterbelts around a large-scale rectangular field. Two belts are fixed at 400 m, the cross belts' distance D was varied and this gives the indicated percentages of mean wind reduction. From Van Eimern *et al* (1964) after Alisov *et al*.

additional growth from the base of the plant, is applied to tall crops of *Ricinus communis* (castor) to prevent lodging. Similarly, clearing away heavy leaves and weak trees in bananas, removing the male part of maize after pollination and stooking early harvested *Sesamum indicum* (simsim) carefully to a constructed fence (so that they are not shaken and seeds are not dispersed) are all examples of labour intensive practices of protecting crops from excessive wind damage. They are further examples of the ingenuity of traditional farmers to manage and manipulate the microclimate.

DISCUSSION

Acland (1971) reports that air-borne dust discourages some of the parasites of scale insects which are found on sisal. Nearby roads or other sources are therefore found to induce localised attacks. Such a cause and effect relationship cannot be obtained by traditional farmers, not even if they happened to associate the dust with the insect attacks. Attention of scientists is needed to reveal the sources and solutions of such problems. But important local work in many pertinent fields of agricultural science is relatively scarce especially outside cash cropping. Harwood (1979) has called detailed classification of environmental factors one of the critical elements in small farm development that is largely or entirely omitted in most current systems of agricultural research and development. Examples given in this paper serve to illustrate this point in regard to crop micrometeorology.

Windbreak research on small-scale around the field barriers is virtually non-existant. On large-scale fields surrounded by shelter some Russian work is relevant (Fig. 2). In addition only the work of Kaiser in Germany, whose ideas are most pertinent to traditional farming conditions (Van Eimern, 1964; 1968) can be mentioned (Fig. 3). Research on wind reduction by scattered trees or patches of bush is non-existant either. Two reasons may be forwarded for explanation. First, the general lack of attention to problems which do not or no longer, through mechanization, occur in larger scale farming. The second is that progress in windbreak and crop damage research was very much enhanced by wind tunnel research. This is most often too sophisticated under Third World conditions and not necessarily effective in understanding wind reduction caused by such scattered roughness elements with their inherent geometrical complexity.

Extensive field studies of air movement and wind have been carried out with relatively simple methods in the period after the war in Europe, Japan, New Zealand and USA. It appears as if such experiments are also the only way to collect relevant information on the conditions prevailing in small scale farming in the tropics. Highly valued support for this work could come from developed countries reorienting parts of their micro-meteorological research. This should include field and wind tunnel experiments on

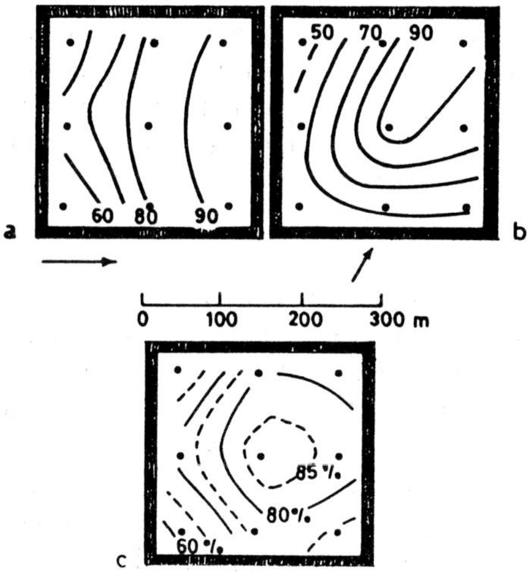

Fig. 3. A rare example of wind speed observations on a quadratic field with barriers all around. Parts a and b give wind speed distribution as a percentage of undisturbed wind for two different wind directions indicated by the arrows. Part c gives the mean wind speed, expressed in the same way but based on the mean frequency of wind direction, with main direction given by the arrow. Dots represent the measuring points. These 4 m high hedges around 7.5 ha fields are of a much larger scale than tropical smallholder fields. From Van Eimern *et al* (1964) after Kaiser.

protection configurations occurring in traditional farming. Efficiency of indigenous practices may be improved and their dissemination may be fostered by such coordinated efforts.

ACKNOWLEDGEMENTS

The author is most grateful to Mr. G. Kalekola for assistance in translating many of more than 100 essays submitted in a contest on examples of microclimate management and manipulation in traditional farming in Tanzania, mentioned in the text. He also thanks participants for their contribution.

REFERENCES

Acland, J.D. (1971). East African Crops. London: Longman.

Battan, L.J. (1983). Weather in Your Life. San Francisco: Freeman & Co.

Beets, W.C. (1982). Multiple Cropping and Tropical Farming Systems. Aldershot: Gower and Boulder: Westview Press.

Casada, J.H., Walton, L.R., Swetnam, L.D. & Duncan, G.A. (1980a). Improving wind resistance of Burley tobacco. University of Kentucky College of Agriculture, Bulletin 717.

Casada, J.H., Walton, L.R. & Swetnam, L.D. (1980b). Wind resistance of Burley tobacco as influenced by depth of plants in soil. Transactions of the American Scoiety of Agricultural Engineers **23**: 903-906.

De Vos, A. (1975). Africa, The Devastated Continent? The Hague: Junk.

Grace, J. (1977). Plant Response to Wind. London: Academic Press.

Greenland, D.J. (1975). Bringing the green revolution to the shifting cultivator. Science **190**: 841-844.

Hurst, G.W. & Rumney, R.P. (1971). Protection of plants against adverse weather. WMO No. 281, Technical Note No. 118, Geneva.

Harrison, P. (1982). The new age of organic farming. New Scientist **24**: 427-429.

Harwood, R.R. (1979). Small Farm Development. Boulder: Westview Press.

ILACO (Ed.), with De Vries, D.A. & Stigter, C.J. (1981). Climate. *In* Agricultural Compendium for Rural Development in the Tropics and Subtropics. Amsterdam: Elsevier.

Jaffe, M.J. (1980). Morphogenetic responses of plants to mechanical stimuli or stress. BioScience **30**: 239-243.

Kangele, P.S. (1981). Diurnal variation of surface winds over a uniform inland area. *In* National Agrometeorological Committee-NewsletterNr. 4, C.J. Stigter (ed.). Dar es Salaam: Directorate of Meteorology.

Levitt, J. (1972). Responses of Plants to Environmental Stresses. New York, London, San Francisco: Academic Press.

Lyles, L. & Allison, B.E. (1981). Equivalent wind-erosion protection from selected crop residues. Transactions of the American Society of Agricultural Engineers **24**: 405-408.

Marshall, J.K. (1967). The effect of shelter on the productivity of grasslands and field crops. Field Crop Abstracts **20**: 1-14.

Rosenberg, N.J. (1974). Microclimate: The Biological Environment. New York: Wiley & Sons.

Stigter, C.J. (1982). Manipulation of microclimate in Tanzanian traditional farming: a preliminary contest report. *In* National Agrometeorological Committee – Newsletter Nr. 7, C,J, Stigter (ed.). Dar es Salaam: Directorate of Meteorology.

Stigter, C.J. (1983). Microclimate management and manipulation in traditional farming. Document 15, item: Land use and agricultural management systems under severe climatic conditions. *In* Final Report, Commission for Agricultural Meteorology, VIII session, W.M.O., Geneva.

Stigter, C.J. (1984a). Traditional use of shade: a method of microclimate manipulation. Archiv für Meteorologie, Geophysik und Bioklimatologie, **34**: 203-210.

Stigter, C.J. (1984b). Examples of mulch use in microclimate management by traditional farmers in Tanzania. Agriculture, Ecosystems and Environment, **10**.

Stigter, C.J., Othieno, C.O. & Mwampaja, A.R. (1984). An interpretation of temperature patterns under mulched tea at Kericho, Kenya. Agricultural Meteorology, **31**: 231-239.

Sturrock, J.W. (1975). Wind effects and their amelioration in crop production. *In* Physiological Aspects of Dryland Farming, U.S. Gupta (ed.). New Delhi: Oxford & IBH Publishing Company. pp. 283-313.

Sturrock, J.W. (1978). Effects of wind. Science **200**: 1263-1264.

Uemura, K., Mihara, Y., Hanya, J. & Omoto, Y. (1974). Meteorological hazards. *In* Agricultural Meteorology of Japan, Y. Mihara (ed.). Tokyo: University of Tokyo Press. pp. 97-125.

Van Eimern, J. (1968). Problems of shelter planning. *In* Agroclimatological Methods. UNESCO-Natural Resources Research Series, No. 7: 157-166.

Van Eimern, J., Karschon, R., Razumova, L.A. & Robertson, G.W. (1964). Windbreaks and shelterbelts. WMO-No. 147 TD. 70, Technical Note No. 59, Geneva.

Waddell, E. (1983). Coping with frosts, governments and disaster experts: some reflections based on a New Guinea experience and a perusal of the relevant literature. *In* Interpretations of Calamity, K. Hewitt (ed.). Boston: Allen & Unwin. pp. 33-43.

Wilken, G.C. (1972). Microclimate management by traditional farmers. Geographical Review **62**: 544-566.

Wilken, G.C. (1977). Integrating forest and small-scale farm systems in Middle America. Agro-Ecosystems **3**: 291-302.

Progress in Biometeorology, 1985, Vol. 2, pp. 155-176

THE EFFECT OF CLIMATE ON PLANT GROWTH AND AGRICULTURE IN THE FALKLAND ISLANDS

J.H. McAdam
Agricultural Botany Research Division,
Department of Agriculture for Northern Ireland
and Department of Agricultural Botany,
Queen's University Belfast,
Newforge Lane,
Belfast, BT9 5PX,
Northern Ireland.

INTRODUCTION

The Falkland Islands lie between Latitude 51° 00' and 52 ° 30'S and Longtitude 57° 40' and 61° 31'W. They are an archipelago comprising 2 main islands, East Falkland (c. 5000 km) and West Falkland (c. 3500 km) together with more than 230 smaller islands varying from 220 km to a few square metres in extent. The total land area of the archipelago is about 10,000-12,000 km (Moore, 1968).

The Islands are generally hilly with several distinct mountain ranges, the highest point being 705 m. The coastline is deeply indented and streams and ponds are numerous throughout the archipelago. The geology of the region has been surveyed in detail (Greenway, 1972) and consists almost entirely of palaeozoic and mesozoic sediments.

The cool moist climate and low soil bacterial activity have combined to promote the formation of fibrous acid peats throughout the Islands. Mineral soils are encountered only in very small areas. In general, natural fertility is low except where penguins or seals have manured the land.

The predominant plant associations found on the Islands are closely related to one another so that the vegetation tends to have a rather monotonous appearance characterised by the absence of any significant tree cover.

Extensive sheep farming is the only industry of any consequence. Characteristic of the economy of the Falkland Islands is the almost total dependence on the income generated from the sale of wool – 88% of the national income in 1981 (Shackleton, 1982). This results in a fragile and often wildly fluctuating economy which is vulnerable to changes in world wool prices – a commodity which has not been buoyant in recent years. The medium term economic outlook is one of a slow decline and this exerts pressure on the need to improve the agricultural output or wool production.

LIMITATIONS TO AGRICULTURE

At present the main limitations to agricultural production are poor individual sheep performance in terms of growth rate and fertility, and high losses in the sheep population. These can be related to the poor nutritional status and unfavourable growth pattern of the natural grasslands, and the adverse climate for plant growth and livestock performance.

It is apparent that the major limiting factor to agriculture is one of inadequate nutrition of the sheep. Year-long, free range, set-stocked pastoral systems operating against the background of highly seasonal pastoral growth inevitably produce a cyclical pattern of nutrition. The dietary energy requirements of ewes are characterised by a minor increase over maintenance requirements in the immediate pre-mating period (April/May) followed by a mid-gestation requirement for maintenance energy intake only. The most important periods in the cycle are the last 6 weeks of pregnancy (September/October) and the lactation period (November-January) over which period the dietary requirements may increase to up to 3 times maintenance levels. In order to achieve acceptable levels of production it is essential that this annual cycle of energy demand is met, especially during the critical late pregnancy and lactation periods.

The isolation of the Islands virtually excludes the widespread use of purchased concentrated feedstuffs and conditions are unsuitable for the growing of fodder and cereal crops. Hence the main resource available is the natural grasslands.

Attempts to improve pasture have been very limited, partly because of the absence of scientifically based information on suitable methods but also because of the generally poor economic returns from such investment.

Information on the overall impact of the unfavourable climate on plant growth and livestock performance will be presented in this paper and the role of shelter outlined.

THE NATURAL GRASSLANDS

Description

The vegetation of the Islands has been classified as an oceanic heath and has been described in detail by Moore (1968) who recognised five formations viz. maritime tussock; oceanic heath; feldmark; fen and bog;

and bush. However most of the Falkland Islands are covered by some of the wide range of vegetation types and communities included in the oceanic heath formation. Within this formation the most common communities are those in which the grass *Cortaderia pilosa* D'Urv Hack is dominant or co-dominant. This species (locally called whitegrass) is the most important component of the native vegetation.

It is a perennial grass which seems to exhibit a range of forms dependent on habitat. In shallow-sided valleys (sheltered areas with a continuous ground water supply) the plant tends to form tussocks whereas on level and undulating ground with poorly drained soils *Cortaderia* produces a uniform cover with a relatively lax growth habit.

It has been estimated that *Cortaderia*-dominated communities comprise, by area, approximately 65% of West Falkland, 53% of East Falkland and 41% of the scattered islands (Davies *et al*, 1971). In view of this dependence on natural grasslands for the dietary needs of the sheep it was considered that basic information on the seasonal growth pattern, above-ground production and quality of *Cortaderia* was fundamental to development of improved sheep grazing systems.

Productivity

Estimates of total annual dry matter production on a range of vegetation types have been obtained using a pre-trimming technique (Milner & Hughes, 1968). The most productive areas are those found in valleys, adjacent to small ponds and lakes and near the coast where mean total above-ground productivity has been estimated at 5,000-6,300 kg $ha^{-1}yr^{-1}$ with the least productive (<1,000 kg ha^{-1} yr^{-1}) being the bog formations, generally found at a high altitude (J.H. McAdam, unpublished report to the Ministry of Overseas Development, 1980). The mean annual production of uniform-cover *Cortaderia*-dominant communities was withing the range 1,250-2,800 kg ha^{-1} yr^{-1}. In another experiment using the difference between standing crops technique to measure net primary productivity, the yield of the lax form of *Cortaderia* was found to be 2,500 kg ha^{-1} yr^{-1} (J.H. McAdam, unpublished data).

Quality

Cortaderia is a tough wiry grass with a high proportion of dead matter in the herbage and hence is of relatively poor value as a source of feed for grazing stock. The quality of pastures dominated by *Cortaderia* is low with organic matter digestibilities in the range 35-45% (J.H. McAdam, unpublished data). Values for comparable pastures in the British Isles would be 50-70%.

Seasonal growth pattern

Information on the seasonal growth pattern of *Cortaderia*-dominant pastures was obtained from measurements on individual leaf extension rates and from staggered cuts overlapped to give 3-weekly estimates of growth.

1. Leaf extension

Individual *Cortaderia* tillers were marked and the length of the youngest green leaf measured at regular (approximately monthly) intervals throughout the 1977/78 season. It can be seen that during the period when shortfalls in the nutrition of the sheep can most severely limit production (Sept-Dec) leaf extension rates were initially low, increasing rapidly to a maximum in late December and declining more slowly in the latter part of the growing season. Narrow-leaved snow tussock *Chionochloa rigida* (Raoul) Zotov. from South Island, New Zealand has been thought by some workers (e.g. Walton, D.W.H., personal communication) to closely resemble *Cortaderia* in many respects including growth habit. Similar data on leaf extension for this species (Mark, 1965) show that the shortfall in spring leaf growth recorded for *Cortaderia* is not found for *Chionochloa* (Fig. 1).

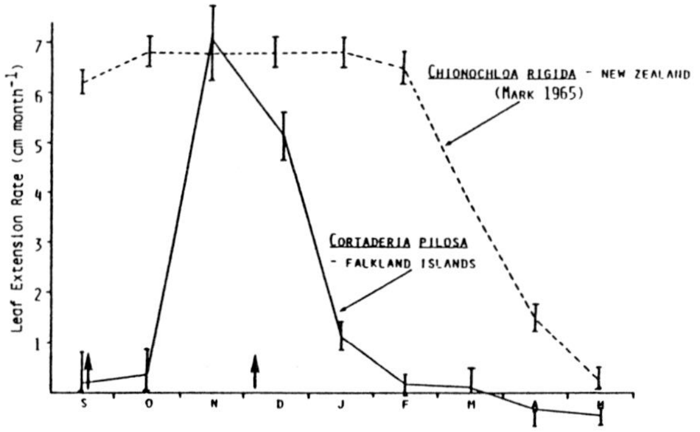

Fig. 1. Rates of extension of the youngest leaves of *Cortaderia pilosa* in the Falkland Islands (1977-78) and *Chionochloa rigida* in New Zealand (1960-61) over the growing season. The vertical bars indicate standard errors and the arrows illustrate the period over which the energy intake demand is greatest.

2. Growth estimates by sequential cutting

On an area of uniform-cover *Cortaderia*, twenty 0.09 m^2 plots were pretrimmed and cut to ground level at 45 day intervals throughout the season and the accumulated herbage dried and weighed. Half the plots were cut 21 days out of sequence enabling overlapping estimates of growth to be made. There are many inherent errors associated with this technique but

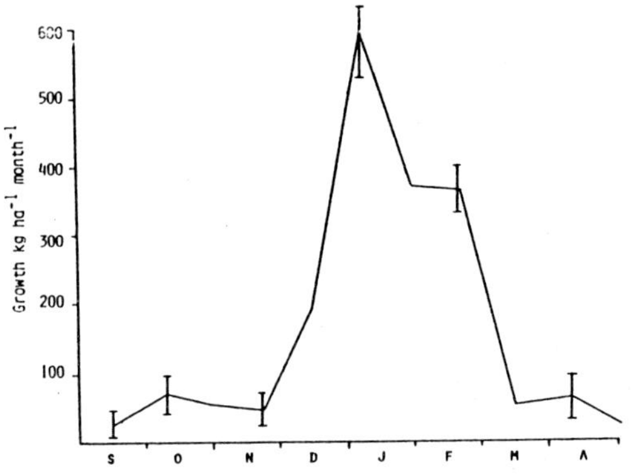

Fig. 2. The growth of *Cortaderia pilosa* cut at 45 day intervals over the growing season following pretrimming. Vertical bars indicate standard errors.

it does provide a rapid and relatively easy estimate of the seasonal pattern of growth although the cumulative yields may overestimate productivity. The absence of early season growth in the 1977/78 season was even more clearly illustrated in this experiment than in the one reported above.

It is clear that from these experiments and from observation, that there are serious shortfalls in early season herbage production at a time when breeding sheep have their maximum energy demand.

In an attempt to overcome this shortfall and to increase the quality of the grazed pasture, sheep farmers have improved areas of natural vegetation by a combination of burning, rotavating and reseeding with Yorkshire fog (*Holcus lanatus*), cocksfoot (*Dactylis glomerata*) and red fescue (*Festuca rubra*). These improvements have met with varying degrees of success because of the large scale of the operations involved, the inherently low soil fertility and the inability to exercise satisfactory grazing control on such areas.

It is clear that the introduced species do, in most situations, provide pasture of increased nutritional value to the sheep but from the evidence available it does not appear that such improved pasture, in the absence of early season application of fertiliser N, significantly improves the poor spring growth situation characteristic of the native pasture.

This has been demonstrated for the 1976/77 season on a *Dactylis glomerata* reseed (Fig. 3) where the seasonal growth pattern was assessed using the sequential cutting technique previously referred to.

Although there was a gradual increase in growth over the early part of the season, the same mid-season peak of production was found as for the *Cortaderia* (Fig. 2). The overall production and quality of this *Dactylis* pasture was, however, much greater than that for the *Cortaderia* and it is apparent that if some modest increases in spring production in such pastures could be obtained, the resultant benefits to current sheep production systems could be substantial. In this context preliminary investigations into the use of early season applications of N fertiliser have shown promise although the economic benefit of such a practice has yet to be demonstrated. The possible role of strategically positioned shelterbelts in this context will be discussed later.

The spring shortfall in production is clearly illustrated when the growth of the same species (*Dactylis glomerata*) in the Falkland Islands is compared with its growth in hill land in New Zealand (Sithamparanathan, 1979, Fig. 3b) and in the British Isles (Alcock *et al*, 1968, Fig. 3c).

THE CLIMATE

In an attempt to explain the shortfall in spring production it is appropriate to consider the various facets of the climate in some detail.

The climate has been described as cool oceanic and is characterised by a narrow temperature range, relatively low spring rainfall and a high frequency of strong winds especially in the spring.

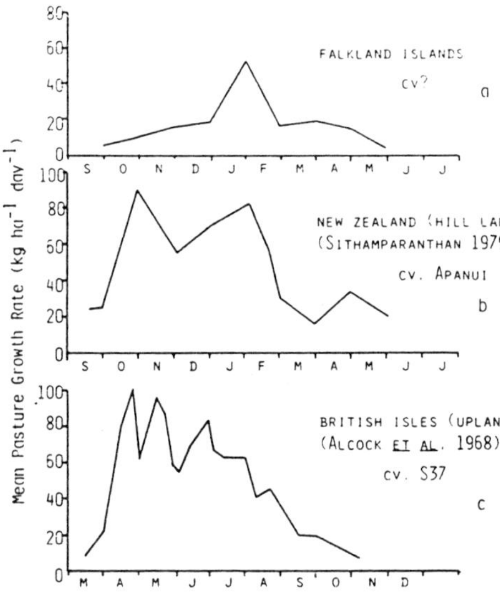

Fig. 3. The growth of cocksfoot (*Dactylis glomerata*) at 3 sites estimated from sequential cutting: a Falkland Islands 1976-77 (cultivar not known), b New Zealand (cv Apanui), c Wales (cv S37).

Temperature

The relatively narrow seasonal variation in temperature is illustrated in Fig. 4 with an indication of extremes of summer and winter temperatures (°C) at Stanley from 1948-1978 gained from the information presented below (Table 1).

Table 1. *Summer and winter temperatures (°C) at Stanley (30 year mean).*

	Mean	Mean Lowest Minimum	Mean Highest Maximum	Mean Minimum	Mean Maximum
January	9.4	1.5	20.6	5.6	13.2
July	2.2	–5.0	7.8	0.2	4.2

The mean winter temperatures are comparable with those experienced in the British Isles but the summer mean is more similar to conditions in Scotland and Western Norway. Air frosts are uncommon in summer but no month is frost-free and ground frosts can occur throughout the year.

Soil temperature data are presented in Fig. 4 and it can be seen that during the spring and autumn period soil and air temperatures are similar. The generally accepted threshold for plant growth is a grass minimum temperature of 5.5 °C (Anslow & Green, 1967). The relation between this temperature and that in the soil at 10 cm has not been clearly defined. However, data from the Falklands show that the grass minimum reaches 5.5 °C concomitant with or slightly later than the 10 cm soil temperature. On this basis the mean duration of the growing season (20 year mean) was 178 days (Grasslands Trials Unit, 1977).

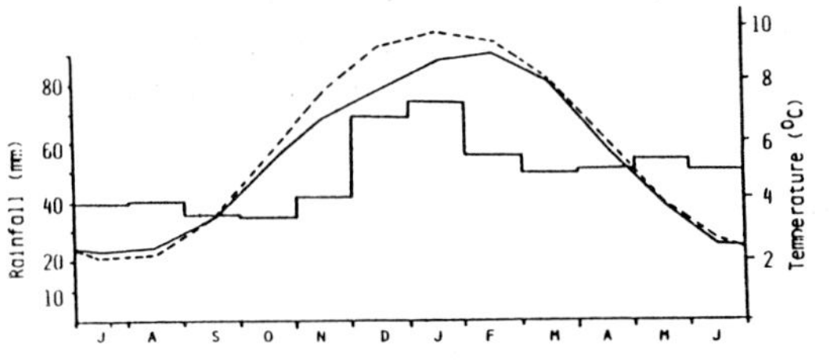

Fig. 4. The mean monthly rainfall (histogram), soil temperature at 10 cm (solid line) and air temperature (broken line). Based on 30 year means at Stanley.

Rainfall

The mean annual precipitation at Stanley during the period 1944-1978 was 640 mm. Rain falls on an average of 140 days in the year and the highest

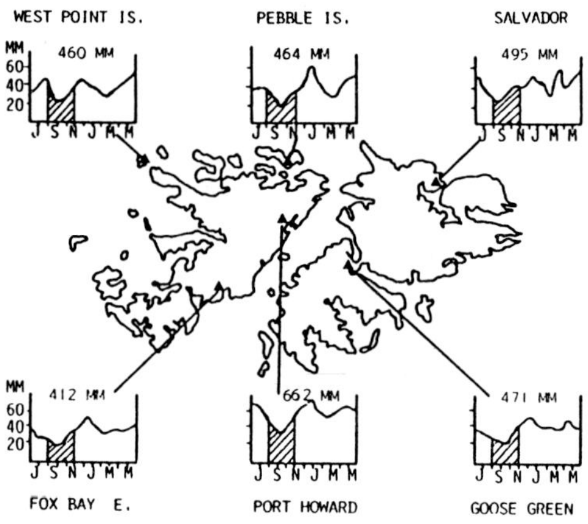

Fig. 5. The seasonal distribution of rainfall from 6 stations (▲) in the Falkland Islands.

rainfall occurs in the summer months with a marked depression in spring (Fig. 4). Stanley is, however, in one of the wetter parts of the Islands, with drier areas generally in the south and west. The marked depression in rainfall during the months of September and October and already referred to for Stanley, is also found at other stations where rainfall records are kept (Fig. 5).

Wind

The relatively strong, and almost continuous, winds are the most notable feature of the climate. The mean hourly windspeed of 8.5 m s^{-1} hardly varies over the year with only a slight increase in spring and early summer (Fig. 6). Over the period 1975-1978 it was found that from mid-September

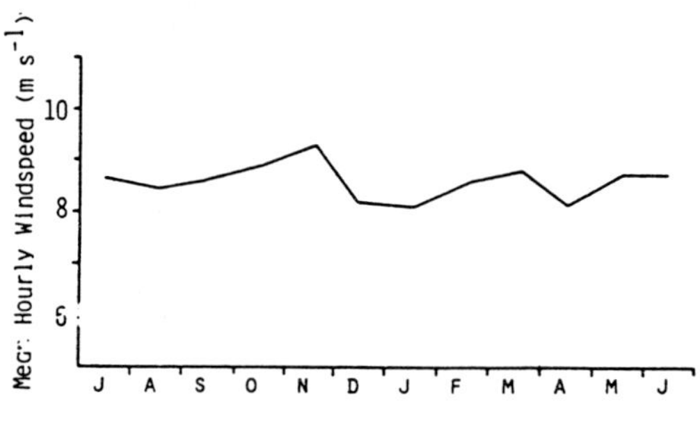

Fig. 6. Mean hourly windspeed at Stanley, East Falkland (1976-79).

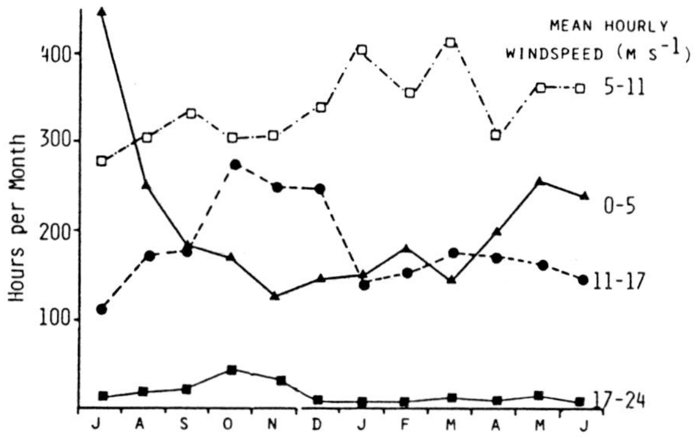

Fig. 7. The number of hours per month when the mean hourly windspeed was 0-5 (▲), 5-11 (□), 11-17 (●) and 17-24 (■) m s^{-1}.

onwards there was a marked decrease in the frequency of low windspeeds (0-5 m s^{-1}) and a corresponding increase in the frequency of winds with speeds greater than 11 m s^{-1} (Fig. 7). In October, for example, on average the windspeed was greater than 11 m s^{-1} for over 300 hours or almost half the month.

Pressure generally decreases steadily from north to south over the Islands, giving gradients for westerly winds and about 50% of the winds are between south-west and north-west with 80% to the west of the north/south line (Pepper, 1954). The frequency of strong winds is greatest from the south-west (Fig. 8). This relative consistency in the direction of strong winds

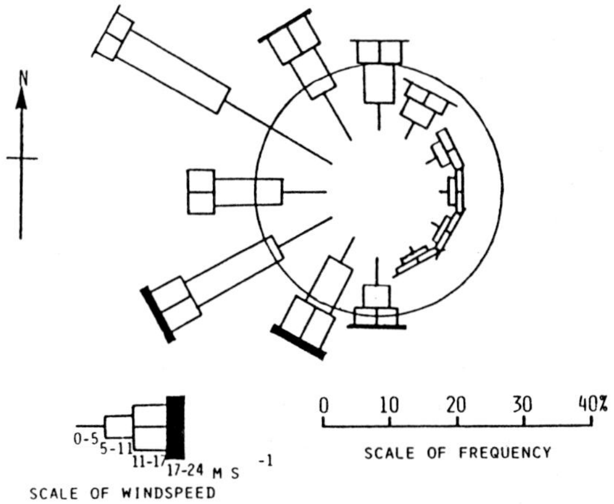

Fig. 8. The mean frequency, speed and direction of the wind at Stanley for the whole year (means 1944-50; Pepper 1954).

has obvious implications in the siting of shelterbelts.

Exposure

It has been suggested that the rate of tatter of standard cotton flags may well be governed by that combination of atmospheric influences which inhibits plant growth (Lines & Howell, 1963) so exposure may be indicated better by tatter flags than by purely wind measuring devices (Savill, 1974). Tatter flags were flown over a two year period at Stanley, Goose Green and Salvador on East Falkland. A full analysis of these data has been published elsewhere (McAdam, 1980). The mean tatter rate over the three sites was 9.76 cm day^{-1}. This was higher than rates recorded for Shetland (9.16 cm day^{-1}), Orkney (7.32 cm day^{-1}) (Lines & Howell, 1963) and exposed forestry sites in N. Ireland (7.96 cm day^{-1}) (Savill, 1974). Tatter rate did increase considerably over the summer (Fig. 9).

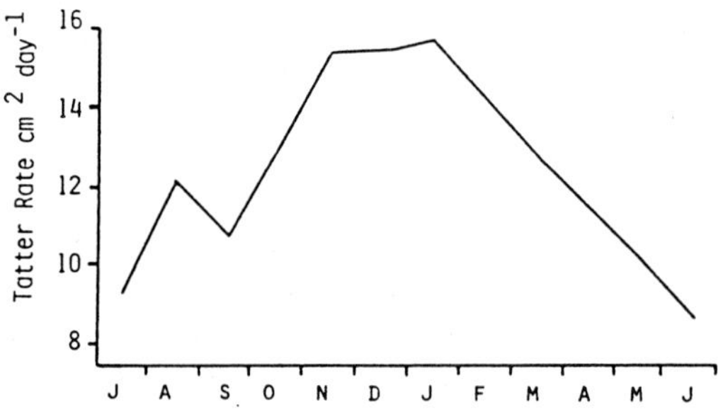

Fig. 9. Mean rate of tatter of standard flags (1976-78) at Stanley, Goose Green and Salvador (East Falkland).

Caution must, however, be exercised in the interpretation of tatter flag data (Rutter, 1966) and an analysis of the factors contributing to tatter at Stanley (McAdam, 1980) indicated that, in contrast to the situation in the British Isles, rainfall was not a primary component of tatter in the Falklands, wind factors accounting for most of the tatter.

Climate and plant growth

Hence it can be seen that overall the climate in spring in the Falklands is characterised by low temperatures, low rainfall and a high frequency of strong winds. It has been suggested by Skottsberg (1913) that the relatively low spring rainfall imposes restrictions on plant growth and Moore (1968) has concluded that the strong winds occurring in spring probably emphasise the effects of low spring precipitation on plant growth.

ANIMAL PRODUCTION AND PERFORMANCE

Herbage quality

It is clear that there is a complex interaction between herbage growth and quality, animal production and the impositions of the various components of the climate. From the results presented previously, at the time of maximum energy demand by the sheep, it is the production of herbage that is limiting.

The theoretical changes in sheep energy requirements over the growing season have been proposed by Davies *et al*, (1971) and are presented in Fig. 10. It can also be seen that although the digestibility of the green leaf component of the *Cortaderia* is increasing in the springtime, concomitant with the increase on energy demand (J.H. McAdam, unpublished data), this component represents only approximately 25% of the above ground herbage, and the overall above-ground herbage digestibility does not reach a peak until late February. There is no doubt that some diet selection will occur although examination of the stomach contents of sheep in September and October revealed that a considerable proportion of the dead and shrub component of the herbage was eaten (R.S. Whitely, personal communication).

Wool production

The effect of this nutritional situation on wool production and sheep performance has not been fully elucidated. However, information on the

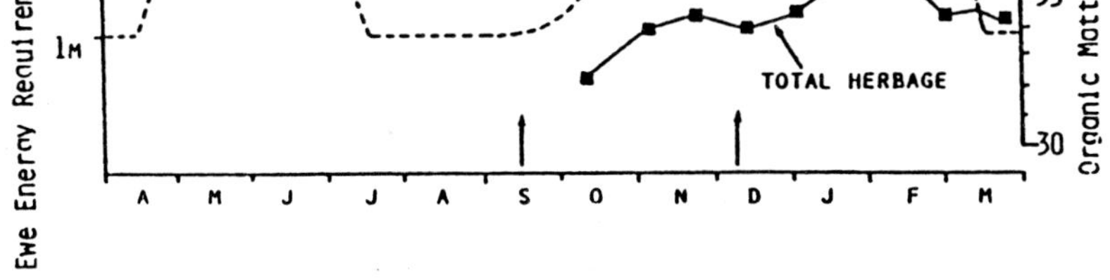

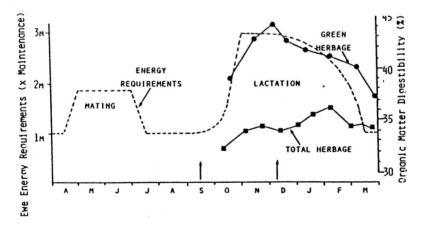

Fig. 10. The variation in theoretical energy requirements of the ewe (as a proportion of maintenance requirements) throughout the year, and organic matter digestibility of *Cortaderia*-dominant herbage in the spring period (actual values, 1977).

seasonal pattern of wool growth has been collected for different classes of sheep over a number of seasons (Grasslands Trials Unit 1979 unpublished report). These data (Fig. 11) indicate that for breeding sheep there is a marked seasonality of production, the pattern of wool growth possibly being

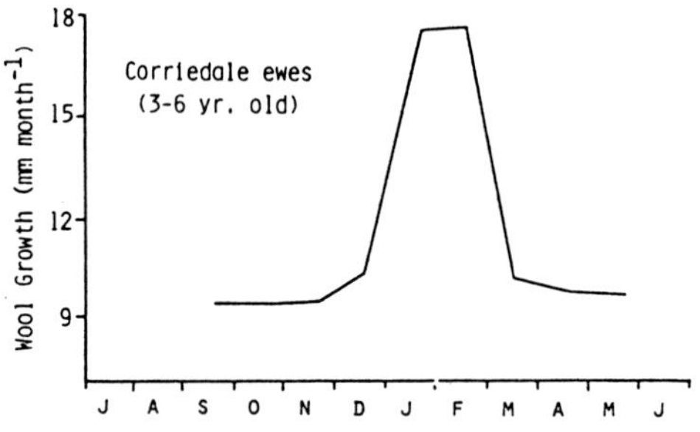

Fig. 11. The seasonal pattern of wool growth of 3-6 year old Corriedale ewes (Salvador, East Falkland 1977-80).

confounded by the additional nutritional demands during late pregnancy and lactation (Ryder & Stephenson, 1968; Coop, 1953). The growth pattern of castrated male sheep (wethers) was somewhat similar with the production spread over a longer period of time (Fig. 12). Comparable data for sheep of the same breed, class and age in New Zealand (Bigham *et al*, 1978) showed a similar trend with some indicatiion that wool growth there commences slightly earlier in the season. It should be noted however, that the wool growth estimates for the Falklands were made on the basis of increase in wool length whereas the estimates reported from the New Zealand sheep were on the basis of weight of wool produced per unit area.

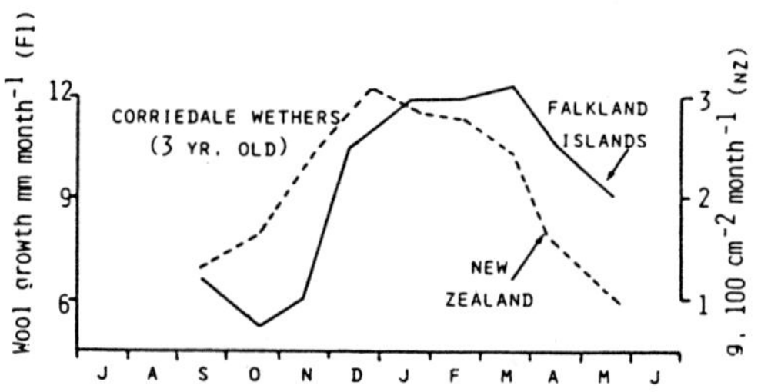

Fig. 12. The seasonal pattern of wool growth of 3 year old Coriedale wethers in the Falkland Islands at Fitzroy, East Falkland 1978-80 (solid line) and in South Island, New Zealand according to Bigham *et al* (1978) (broken line).

The relationship between the two methods of measurement, although unclear, is unlikely to vary appreciably over comparable growth periods.

It has been found that poor nutrition in the early post-natal life of the lamb retards the growth and development of later secondary follicles and permanently limits their production (Doney & Smith, 1964). It has also been shown that improved nutrition during the lactation period can result in increased wool production (Doney, 1964). Coop (1953) has shown that the level of nutrition during winter pregnancy is more important than in lactation. There are breed differences in these patterns and it has been found that the Corriedale (the principal breed of sheep on the Falkland Islands) behaves more like a Merino (Ferguson *et al*, 1949) than a Scottish Blackface, the latter having a distinct cyclic pattern of wool production (Doney, 1966). In the Merino, wool production appears to be temperature and nutrition dependent (Ryder & Stephenson, 1968) and recently Rowe (1982) has shown a direct relationship between annual wool production and pasture dry matter production. Hence, although positive information is lacking, it is reasonable to assume that the nutritional deficits imposed by poor pasture production, itself a function of the climate, may be limiting production.

Stock losses

Ferguson (1980) has reported the annual loss of adult sheep as 10% and Davies *et al* (1971) report overall lamb losses in the first year as 20% (Fig. 13). It is apparent that lamb losses in the early post natal period (Sept/Oct)

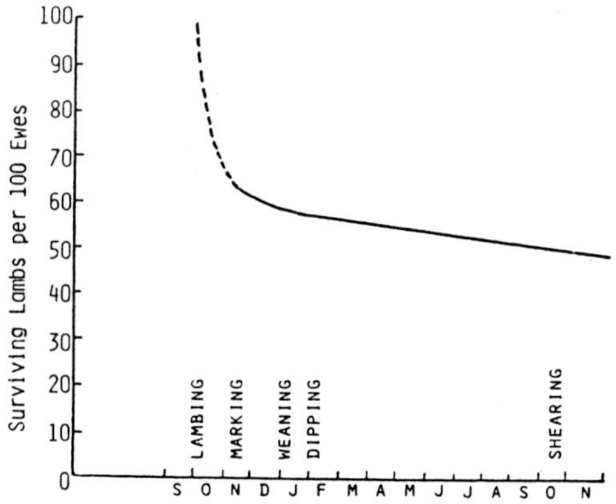

Fig. 13. The survival of lambs born in 1967 over all farms in the Falkland Islands (from Davies *et al*, 1971).

must be substantial although detailed information is not yet available.

Live weight gains

These have been well documented in the Grasslands Trials Unit report (1979) and by Ferguson (1980), and show a reduction in winter followed by a spring increase.

FACTORS AFFECTING HERBAGE GROWTH AND ANIMAL PERFORMANCE

In an attempt to look at the relationship between some of the climatic, plant and animal production variables a correlation matrix was constructed and a series of multiple regressions was carried out selecting up to four variables in any equation. The data are summarised from a series of experiments carried out from 1976-1978, some of which are reported in this paper and most of which are detailed elsewhere (Grasslands Trials Unit, 1979; McAdam, 1980, unpublished reports to the Overseas Development Administration).

The equations are not particularly useful for predictive purposes and have not been presented. However, it is of interest to note which variables are selected when herbage growth and animal performance parameters are regressed against a number of climatic variables:

Plant growth parameters
- Leaf extension (extension rates of the youngest leaf – Fig. 1)
- Leaf dieback (dieback from the tip of the youngest leaf)
- RGR (Relative growth rate of *Cortaderia* – dominant pasture)

Animal performance parameters
- Wool growth (as in Figs. 11 and 12)
- LWG (Live weight gains of dry sheep from Ferguson (1980) and GTU (1979, unpublished report))
- Lamb survival (From GTU, 1979; unpublished data and Davies *et al*, 1971)

Climatic variables
- Sun (mean daily hours sunshine)
- Rain (mean monthly rainfall)
- Mean max, mean min, highest max, lowest min, mean temp. (based on mean daily air temperatures)
- Soil 10 cm (soil temperature at 10 cm depth)
- Wind (mean hourly wind speed)
- Hrs 0-5, Hrs 5-11, Hrs 11-17, Hrs 17-24 (number of hours when the mean hourly windspeed was 0-5, 5-11, 11-17, and 17-24 m s^{-1})
- Gusts 17-24 (no. of hours when the maximum gust recorded was 17-24 m s^{-1})
- Tatter (rate of tatter of standard flags)

In the data presented below the correlation coefficient (*r*) between each variable and parameter and the percentage of the variation accounted for by including the extra variables in the equations are presented (Table 2).

Higher rates of leaf extension were positively correlated with higher levels of sunshine and rainfall, supporting the hypothesis that low spring rainfall may be restricting extension. When more variables were included in the equations, those selected were related to soil and air temperatures.

As would be expected variables related to strong gusting winds accounted for most of the variation in leaf dieback rates. Dieback increased steadily throughout the season when winds were strongest and this analysis would support the view that frequent strong winds are contributing to slow pasture growth and the high proportion of dead material present on *Cortaderia* plants.

The high correlation of RGR with rainfall and high negative correlations with strong wind variables would tend to support the view of Moore (1968) that these factors are restricting grass growth in the spring.

With wool growth in both classes of sheep no overall trends emerged (Table 3). However, temperature and rainfall variables accounted for most of the variation in the dry sheep or wethers. This would tend to explain the different cyclic patterns of production for the two classes of sheep noted earlier (Figs. 11 and 12).

The period of greatest liveweight gain occurred in midsummer when the frequency of strong winds had decreased but the correlations are not high.

As would be expected high death rates in lambs were correlated with low temperatures and strong winds, the highest rates occurring in spring when

Table 2. *Plant growth parameters, variables in equation and % variation accounted for. The correlation coefficient (r) is in parentheses*

Parameter	Variables in Equation	% variation accounted for
a. Leaf extension	1. Sun (0.76)	51
	2. Rain (0.64)+Sun	58
	3. Rain+Mean Min (0.31)+Soil 10 cm (0.50)	87
	4. Rain+Sun+Mean Min+Hrs 11-17 (–0.74)	97
b. Leaf dieback	1. Sun (0.77)	56
	2. Gusts 17-24 (0.77)+Tatter (0.71)	66
	3. Tatter+Hrs 11-17 (0.75)+Gusts (17-24)	74
	4. Sun+Tatter+Hrs 11-17+Hrs 17-24 (0.22)	99
c. Relative Growth Rate	1. Rain (0.65)	33
	2. Rain+Hrs 11-17 (–0.25)	41
	3. Rain+Hrs 11-17+Hrs 17-24 (–0.52)	59
	4. Rain+Mean Min (0.25)+Hrs 11-17+Hrs 17-24	72

wind-chill factors are dangerously high.

Hence it can be seen that there are complex interactions between climate, pasture growth and animal performance. Some of these interactions may be direct and others indirect but the overall effect is to reduce the efficiency and output of the system.

SHELTER

From the above analyses it can be seen that there is a case for the provision of some form of shelter from the wind to have either a direct effect – in terms of enhanced stock growth and lamb survival – or an indirect effect in terms of improved sheep nutrition.

The effect of shelter on pasture growth

In a preliminary investigation a comparison was made between the growth of 'unimproved' (*Cortaderia* dominant) pasture and 'improved'

Table 3. *Animal performance parameters. The correlation coefficient (r) is in parentheses.*

Parameter	Variables in Equation	% variation accounted for
a. Wool growth in dry sheep	1. Mean Min (0.64)	34
	2. Tatter (–0.50)+Rain (0.59)	61
	3. Rain+Mean Max (0.57)+Wind (–0.10)	79
	4. Rain+Tatter+Hrs 17-24 (–0.55)+Wind	96
b. Wool growth in wethers	1. Hrs 17-24 (–0.76)	52
	2. Tatter (–0.22)+Mean Min (0.75)	82
	3. Tatter+Mean Min+Hrs 17-24 (–0.76)	90
	4. Mean Max (0.49)+Hrs 17-24+Wind (–0.02) +Mean Temp (0.65)	96
c. Liveweight gain, ewes and wethers	1. Gusts 17-24 (–0.61)	30
	2. Tatter (0.05)+Gusts 17-24	41
	3. Gusts 17-24+Mean Wind (0.22)	52
	4. Tatter+Gusts 17-24+Mean Wind+Lowest Min (0.65)	67
d. Lamb Survival	1. Lowest Min (–0.69)	41
	2. Tatter (–0.19)+Lowest Min	48
	3. Tatter+Lowest Min+Mean Min (–0.61)	48
	4. Tatter+Lowest Min+Hrs 11-17 (–0.21) +Hrs 17-24 (–0.41)	59

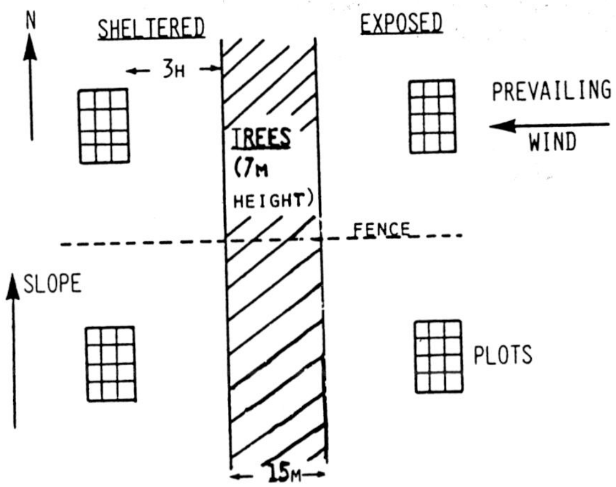

Fig. 14. Site details of an experiment to compare the growth of 'unimproved' (north of fence) and 'improved' (south of fence) pasture in a 'sheltered' and an 'exposed' situation to the west of Stanley, East Falkland.

(*Agrostis magellanica, Holcus lanatus, Poa pratensis* dominant) pasture in a 'sheltered' and an 'exposed' situation (Fig. 14). The shelter was provided by a 15 m wide belt of *Cupressus macrocarpa* (7 m tall) situated immediately behind Government House, Stanley. Plots (0.25 m^2) protected from grazing were pretrimmed in early September 1977 and 4 different plots cut on each of 3 occasions throughout the growing season. The plots to the north of the fence were in native, mostly ungrazed *Cortaderia* and those south of the fence were on lightly grazed pasture which had probably been 'improved' many years previously.

The plots were protected from grazing and the dry matter yields (cumulative totals) at each harvest presented in Table 4 below.

Overall yields were greater on the 'improved' than the 'unimproved' and on the 'sheltered' than the 'exposed' plots on all occasions. On the 'improved' plots in spring, the yield was twice as great in the 'sheltered' as in the 'exposed' situations. These results must be treated with some caution as confounding effects such as the higher level of soil fertility in the plots leeward of the trees due to stock habitually sheltering there may have some enhancement effect. However, from experience from other sites of the effects of varying soil fertility on yield, it was thought unlikely that all of the observed differences could be accounted for by factors such as fertility.

Table 4. *Dry matter yields (kg ha^{-1}) at each harvest (each value is the mean of four plots) from unimproved and improved pasture in sheltered and exposed situations.*

		Sheltered	Exposed
'Unimproved'	21 Nov	330	210
	27 Jan	2610	1820
	28 April	2960	1790
'Improved'	21 Nov	660	310
	27 Jan	3280	2620
	28 April	4140	3410

Enhancement of spring growth in sheltered positions has not been clearly demonstrated experimentally. Jensen (1954) found that shelter increased perennial ryegrass growth by 40 g m^{-2} at the end of June in Denmark and Alcock *et al* (1976) found that shelter did not enhance growth of the same species in upland swards in N. Wales up until the end of May.

Russell and Grace (1979) found no effect of shelter on dry matter production of grass in southern Scotland in spring although their sites did not experience as strong and frequent spring winds as this site.

Grace and Pitcairn (1980) reported a 20% decrease in leaf extension in wind speeds up to 7.4 m s^{-1} and concluded that moisture stress and shortage of available phosphate may exacerbate the situation. The detrimental influence of mechanical stimulation i.e. shaking has been stressed by Grace and Pitcairn (1980) and Grace (1981). Hence, it seems likely that in the Falkland Islands some form of a combination hypothesis may account for the overall restrictions placed by the climate on plant growth. The strong gusting spring winds (mechanical shaking), low spring rainfall (moisture stress), low soil temperatures and low available soil P levels (soil nutrition) (King *et al*, 1968; Davies *et al*, 1971) are all probably responsible to some degree for the observed limitations to spring growth. In view of the relative harshness of the climate and very low levels of available soil nutrients (King *et al*, 1968) it is likely that the effects of shelter on spring grass growth would be more pronounced in the Falkland Islands than in any of the experiments referred to above.

The provision of shelter

The current status of forestry in the Islands has been reviewed by McAdam (1982) who has concluded that the sparsity of trees is due to a combination of undeveloped soils (Stewart, 1982) and lack of advice and encouragement in the past. Observations and trials carried out in the Islands

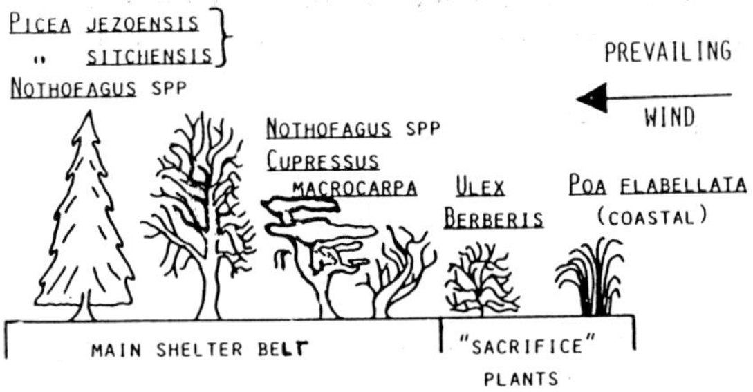

Fig. 15. Proposed structure and species composition of shelter belts for the Falkland Islands.

to date (McAdam, 1982) indicate that trees can be grown moderately successfully where the soil is cultivated and nutrient deficiencies corrected. It is clear that tree growth will never be substantial on any but a few sites; however, the integration of properly designed, protected and strategically placed shelterbelts into the existing farming structure could be of extreme benefit.

In planning shelter belts it is desirable that 'sacrifice' rows of hardy trees or shrubs should be planted at the edge of the plantation (Fig. 15). Suitable species might include Gorse (*Ulex europaeus*), *Berberis*, Tussac grass (*Poa flabellata* – coastal) or some of the smaller, hardier *Nothofagus* species . More upright, hardy trees such as *Pinus mugo, Pinus contorta* and *Cupressus macrocarpa* could be planted next to the leeward with possibly *Picea* (such as are found in the most successful plantation on the Islands, at Hill Cove) and *Nothofagus* species forming the main body of the shelter belt. The prevailing winds are consistently westerly and the siting of shelter belts to allow for maximum effect should present no problems. If the proposed development of the agricultural industry of the Islands is to proceed in a manner which would involve some fragmenting of the large farms and limited intensification and fodder cropping on a small scale (Shackleton, 1982), the prospects for forestry would seem to be favourable.

ACKNOWLEDGEMENTS

The Meteorological Office, Stanley, provided most of the meteorological data reported and Mr. Peter Maitland, Grasslands Trials Unit, Stanley, kindly forwarded records of wool growth. The author is grateful for assistance in the prepartation of figures for the conference by members of the Agricultural Botany Research Division, Department of Agriculture for Northern Ireland, which funded the author's attendance at the conference. The experimental work reported here was carried out while the author was a member of the Grasslands Trials Unit which is financed by the Overseas Development Administration.

REFERENCES

Alcock, M.B., Lovett, J.V. & Machin, D. (1968). Techniques used in the study of the influence of environment on primary pasture production in hill and lowland habitats. In The Measurement of Environmental Factors in Terrestrial Ecology (R.M. Wadsworth, ed.). British Ecological Society Symposium 8: 191-203. Oxford: Blackwell Scientific Publications.

Alcock, M.B., Harvey, G. & Tindsley, S.F. (1976). The effect of shelter on pasture production on hill land in north Wales. Proceedings of the 4th Shelter Symposium, 88-104. Cambridge: MAFF.

Anslow, R.C. & Green, J.O. (1967). The seasonal growth of pasture plants. Journal of Agricultural Science, **68**: 109-122.

Bigham, M.L., Sumner, R.M.W. & Elliott, K.H. (1978). Seasonal wool production of Romney, Coopworth, Perendale, Cheviot and Corriedale wethers. New Zealand Journal of Agricultural Research, **21**: 377-382.

Coop, I.E. (1953). Wool growth as affected by nutrition and by climatic factors. Journal of Agricultural Science, **43**: 436-472.

Davies, T.H., Dickson, I.A., McCrea, C.T., Mead, H. & Williams, W.W. (1971). The sheep and cattle industries of the Falkland Islands. London: Overseas Development Administration.

Doney, J.M. (1964). The fleece of the Scottish Blackface sheep IV. The effects of pregnancy, lactation and nutrition on seasonal wool production. Journal of Agricultural Science, **62**: 59-66.

Doney, J.M. (1966). Breed differences in response of wool growth to annual nutritional and climatic cycles. Journal of Agricultural Science, **67**: 25-30.

Doney, J.M. & Smith, W.F. (1964). Modification of fleece development in Blackface sheep by variation in pre- and post-natal nutrition. Animal Production, **6**: 155-167.

Ferguson, J.A. (1980). Grasslands Trials Unit, Falkland Islands. Stanley: Grasslands Trials Unit.

Ferguson, K.A., Carter, M.B. & Hardy, M.H. (1949). Studies of comparative fleece growth in sheep. 1. The quantitative nature of inherent differences in wool growth rate. Australian Journal of Scientific

Research Series B, **2**: 42-81.

Grace, J. (1981). Some effects of wind on plants. In Plants and their Atmospheric Environment. British Ecological Society Symposium (J. Grace, E.D. Ford & P.G. Jarvis eds.) 21: 31-56. Oxford: Blackwell Scientific Publications.

Grace, J. & Pitcairn, C. (1981). The effect of wind on grass growth – a review of recent work. In Plant Physiology and Herbage Production. British Grassland Society Occasional Symposium (C.E. Wright ed.) 13: 125-130. Hurley: British Grassland Society.

Grasslands Trials Unit (1977). Meteorological data of Agricultural significance from Port Stanley, Falkland Islands. Stanley: Meteorological Office and Grasslands Trials Unit.

Greenway, M.E. (1972). The Geology of the Falkland Islands. British Antarctic Survey, Scientific Reports No 72. London: Natural Environment Research Council.

Jensen, M. (1954). Shelter Effect. Copenhagen: Danish Technical Press.

King, R.B., Lang, D.M. & Blair-Rains, A. (1969). Land Systems Analysis of the Falkland Islands with notes on the soils and grasslands. Miscellaneous report of the Land Resources Division, Directorate of Overseas Survey No. 70. London: Overseas Development Administration.

Lines, R. & Howell, R.S. (1963). The use of flags to estimate the relative exposure of trial plantations. Forestry Commission, Forestry Record No. 51. London: HMSO.

McAdam, J.H. (1980). Tatter flags and climate in the Falkland Islands. Weather, **35**: 321-327.

McAdam, J.H. (1982). Recent tree planting trials and the status of forestry in the Falkland Islands. Commonwealth Forestry Review, **61**: 259-267.

Mark, A.F. (1965). The environment and growth rate of narrow-leaved snow tussock, *Chionochloa rigida*, in Otago. New Zealand Journal of Botany, **3**: 73-103.

Milner, C. & Hughes, R.E. (1968). Methods for the Measurement of Primary Production in Grasslands. International Biological Programme Handbook No. 6. Oxford: Blackwell Scientific Publications.

Moore, D.M. (1968). The vascular flora of the Falkland Islands. British Antarctic Survey, Scientific Reports No. 60. London: Natural Environment Research Council.

Pepper, J. (1954). The meteorology of the Falkland Islands and Dependencies 1944-1950. London: Falkland Islands and Dependencies Meteorological Service.

Rowe, B.A. (1982). A relation between wool production per animal and annual pasture dry matter production per animal. Australian Journal of Agricultural Research, **33**: 705-709.

Russell, G. & Grace, J. (1979). The effect of shelter on the yield of grasses in southern Scotland. Journal of Applied Ecology, **16**: 319-330.

Rutter, N. (1968). Tattering of flags at different sites in relation to wind and weather. Agricultural Meteorology, *3*: 153-165.

Ryder, M.L. & Stephenson, S.K. (1968). Wool Growth. London: Academic Press.

Savill, P.S. (1974). Assessment of the economic limit of plantability. Irish Forestry, **31**: 22-35.

Shackleton, Lord E.A.A.S. (1982). The Economic Development of the Falkland Islands. London: HMSO.

Sithamparanathan, J. (1979). Seasonal growth patterns of herbage species on high rainfall hill country in northern North Island. New Zealand Journal of Experimental Agriculture, **7**: 157-162.

Skottsberg, C.J.F. (1913). A Botanical Survey of the Falkland Islands. Kungliga Svenska vetenskapsakademiens handlingar, **50**: 1-129.

Stewart, P.J. (1982). Trees for the Falkland Islands. Commonwealth Forestry Review, **61**: 219-225.